Boeing, the 737 MAX Crisis and Aviation Safety

Boeing was once a symbol of engineering excellence, a company driven by the innovation and precision of its engineers. However, in the late 20th century, a shift in priorities – focusing on shareholder returns over passenger safety – marked a turning point that culminated in the tragic crashes of two 737 MAX aircraft and the global grounding of the fleet.

This book explores how the pursuit of profit compromised safety principles, neglected human factors, and led to critical design failures that no one at Boeing had a complete view of. It examines the specifics of the Boeing crisis and looks at it in a broader industrial and economic context. This book draws from the author's extensive research on other global industrial accidents, such as the Deepwater Horizon disaster, and highlights the dangers of profit-first decision-making, the necessity of robust safety systems, and the need for legal systems to respond effectively to corporate malfeasance. It demonstrates how neglecting safety as a top priority can lead to massive reputational damage in the airline industry. Written in an accessible style, this book brings clarity to complex issues, and readers will gain practicable insights into how companies, regulators, and stakeholders can better manage the risks of catastrophic failures.

Boeing, the 737 MAX Crisis and Aviation Safety: The Perils of Profit-Driven Engineering is an essential read for professionals in the fields of human factors, safety, aeronautical/aerospace engineering, and law and corporate governance.

Andrew Hopkins is Emeritus Professor of Sociology at the Australian National University in Canberra, Australia. He holds BSc and MA degrees in Sociology from the Australian National University and a PhD degree in Sociology from the University of Connecticut, USA. Andrew is an honorary fellow of the Institution of Chemical Engineers and a life member of the Australian Institute of Health & Safety, where he received a lifetime achievement award. He served as an expert witness at the Royal Commission into the 1998 Exxon gas plant explosion and as a consultant to the U.S. Chemical Safety Board during its inquiries into the 2005 BP Texas City Refinery disaster and the 2010 BP Gulf of Mexico oil spill. Andrew has authored multiple books on industrial and mining disasters, with his titles selling over 100,000 copies.

Boeing, the 737 MAX Crisis and Aviation Safety

The Perils of Profit-Driven Engineering

Andrew Hopkins

CRC Press
Taylor & Francis Group
Boca Raton London New York

CRC Press is an imprint of the
Taylor & Francis Group, an **informa** business

First edition published 2025
by CRC Press
2385 NW Executive Center Drive, Suite 320, Boca Raton FL 33431

and by CRC Press
4 Park Square, Milton Park, Abingdon, Oxon, OX14 4RN

CRC Press is an imprint of Taylor & Francis Group, LLC

Library of Congress Cataloging-in-Publication Data
Names: Hopkins, Andrew, 1945- author
Title: Boeing, the 737 MAX crisis and aviation safety : the perils of
profit-driven engineering / Andrew Hopkins.
Description: First edition. | Boca Raton, FL : CRC Press, 2026. |
Includes bibliographical references and index.
Identifiers: LCCN 2025006276 (print) | LCCN 2025006277 (ebook) |
ISBN 9781041046479 hardback | ISBN 9781041041313 paperback |
ISBN 9781003629269 ebook
Subjects: LCSH: Boeing Company—Management |
Boeing 737 (Jet transport)—Accidents |
Corporations—Moral and ethical aspects—United States |
Aircraft industry—United States—Management |
Aeronautics—Safety measures | Airplanes--Design and construction
Classification: LCC TL686.B65 H66 2026 (print) | LCC TL686.B65 (ebook) |
DDC 629.130028/9—dc23/eng/20250326
LC record available at https://lccn.loc.gov/2025006276
LC ebook record available at https://lccn.loc.gov/2025006277

ISBN: 9781041046479 (hbk)
ISBN: 9781041041313 (pbk)
ISBN: 9781003629269 (ebk)

DOI: 10.1201/9781003629269

Typeset in Times
by codeMantra

Contents

About the Author

Andrew Hopkins is Emeritus Professor of Sociology at Australian National University, Canberra.

Contact: andrew.hopkins@anu.edu.au

Andrew was a consultant to the U.S. Chemical Safety Board in its investigation of the BP Texas City Refinery disaster of 2005 and also for its investigation into the BP Gulf of Mexico oil spill of 2010. He was an expert witness at the Royal Commission into the 1998 Exxon gas plant explosion near Melbourne. He has written books about these accidents as well as books on mining disasters. Over 100,000 copies sold.

He has been involved in reviews of Work Health and Safety regulations and regulators and has done consultancy work for major companies in the mining, petroleum, chemical and electrical industries, as well as for defence.

His educational qualifications and professional achievements comprise:

- BSc and MA (Sociology) from Australian National University and PhD (Sociology) from the University of Connecticut.
- Independent member of Air Force Board of Inquiry into poisoning of F111 maintenance workers. Author of the Board's report.
- Winner of the 2008 European Process Safety Centre safety award, becoming the first individual outside Europe to achieve this honour.
- Honorary fellow of the Institution of Chemical Engineers in recognition of his "outstanding contributions to process safety and to the analysis of process-safety–related incidents".
- Life member of the Australian Institute of Health & Safety. Recipient of award for "lifetime achievement".
- Officer of the Order of Australia (AO) in recognition of his "distinguished service to industrial safety and accident analysis".
- Former member of the advisory board of NOPSEMA – the Australian National Offshore Petroleum Safety and Environmental Management Authority.
- Member of expert panel that drafted the Global Industry Standard on Tailings Management.

His books include:

Making Safety Work (1995)
Managing Major Hazards: The Moura Mine Disaster (1999)
Lessons from Longford: The Esso Gas Plant Explosion (2000)
Lessons from Longford: The Trial (2002)
Safety, Culture and Risk (2005)
Lessons from Gretley: Mindful Leadership and the Law (2007)
Learning from High Reliability Organisations (2009). Edited

Failure to Learn: the BP Texas City Refinery Disaster (2008)
Disastrous Decisions: Human and Organisational Causes of the Gulf of Mexico Blowout (2012)
Nightmare Pipeline Failures: Fantasy Planning, Black Swans and Integrity Management (2014) with Jan Hayes
Risky Rewards: The Effect of Company Bonuses on Safety (2015) with Sarah Maslen
Quiet Outrage: The Way of a Sociologist (2016)
Organising for Safety: How Structure Creates Culture (2019)
Credibility Crisis: Brumadinho and the Politics of Mining Industry Reform (2021) with Deanna Kemp
Sacrificing Safety: Lessons for Chief Executives (2022)

Acknowledgements

I acknowledge my debt to Peter Robison whose monumental book, *Flying Blind*, published in 2021, provides an exhaustive account of events to that time. I have drawn freely on this source. Since its publication, much has happened and will continue to happen. It is a never-ending story.

I would like to extend my thanks to Dominic Gates, the prize-winning *Seattle Times* journalist, who has followed this story for many years and continues to do so.

I am grateful to Heather McGregor, Bernard Davis, Tamar Hopkins, and Anthony Hopkins for meticulous editing and comments. I deeply appreciate Heather's enduring support and her role as a sounding board, always available to me.

1 Introduction

Shortly after take-off, the Boeing 737 MAX seemed to develop a mind of its own. Its nose dropped sharply, and it descended at a steep angle towards the sea below, as if intent on self-destruction. The pilots had no idea what had happened and struggled to bring the aircraft back to level flight. But the plane over-rode their efforts and insisted on completing its death dive, killing all on board. Four months later, a second 737 MAX took over from its pilots, in exactly the same way, and drove the aircraft into the ground, again killing all on board. A total of 346 people died in these crashes.

The aircraft were flown by airlines in Indonesia and in Ethiopia and the crashes occurred in 2018 and 2019. The 737 MAX was a brand-new version of the tried-and-true 737 that had first gone into service in 1967.

Arguably, the new aircraft should have been grounded after the first crash. It took the second crash to convince regulators around the world to ground hundreds of these new planes that were already in service.

In short order, Boeing was found to have made a series of questionable engineering decisions in designing the new version of the 737. It had also misled its airline customers in an effort to get the aircraft into service as quickly as possible. It had even misled the regulator about the new aircraft.

To cap it off, early in 2024, a door-sized panel on an Alaskan Airlines 737 MAX blew out in flight. The problem? The panel was missing the bolts that were supposed to hold it securely in place. This had been overlooked by Boeing's assembly line quality control process – a quality escape, to use Boeing's jargon. No one was injured, but the potential for another catastrophic event was obvious to all. The ensuing investigations revealed a pattern of manufacturing quality failures. Some of Boeing's top executives – the CEO, the Chief Financial Officer, and the head of Boeing's civilian aircraft business – saw this as a turning point and announced their departures.

Boeing had long been a highly respected aircraft manufacturer. This book examines how Boeing lost its safety focus and identifies some of the tentative steps since taken to re-prioritise safety. I have been particularly interested in what Boeing has learnt from its failures and to what extent it has managed to embed that learning into its organisational structure and functioning. Boeing's attempt to re-construct itself is a work in progress, so this book can only be a snapshot along the way.

The organisational factors that contributed to Boeing's failure have the potential to lead to disaster in any corporation operating in a hazardous industry. All hazardous industries, therefore, have much to learn from Boeing's experience.

This analysis is informed by a study of major accidents more generally. I hope readers already familiar with the Boeing story will find this wider context illuminating.

DOI: 10.1201/9781003629269-1

WHY *THIS* BOOK?

The enormous interest generated by Boeing's failures makes this a particularly appropriate case study from which to learn, so let us step back for a moment and identify a major source of this interest.

First, consider this contrasting case.[1] In 2015 and again in 2019, two different mine sites in Brazil experienced very similar mining accidents. Both mines were owned by global mining giants – in the first case, BHP and Vale, jointly, and in the second, Vale, alone. Both accidents involved the failure of tailings dams – earth dams that serve to impound huge quantities of semi-liquid mining waste material – tailings. The first of these failures killed 19 people and polluted the river downstream for many hundreds of kilometres to the sea, destroying the livelihoods of vast numbers of people. The second did less environmental damage, but killed 272 people. The companies are being sued for billions of dollars. These two failures shook the mining industry globally and led to the creation of a new global standard for the management of tailings dams. They also sent shock waves through Brazilian society. But outside of these two contexts, they barely registered. A tiny proportion of people around the world are personally at risk from tailings dam failures and most people have no idea of what a tailings dam is, let alone what can happen if one fails. A book written about tailings dam failures is therefore unlikely to gain much attention outside the mining industry. I know, because I have written one.

Compare this with aircraft safety. Air travel is extremely widespread in affluent societies. As a result, nearly all of us can readily identify with the victims of an aircraft crash that kills everyone on board; that is, we can readily imagine it happening to us. The fact that it may have happened in far-off Ethiopia or Indonesia does not alter this identification, because we know we travel on very similar aircraft produced by the same manufacturer, and so, may be just as much at risk as victims in faraway lands. Even more to the point, the most powerful people in modern societies – corporate executives, politicians, senior public servants, and journalists – take flights far more often than most people. They are likely, therefore, to be particularly attuned to stories of malfeasance by aircraft manufacturers and airlines. A sociologist once coined the term "victim constituency"[2] – those who can imagine themselves in the victims' shoes. Applying that idea here, there is a much larger and more powerful victim constituency for aircraft crashes than for tailings dam failures. Accordingly, the potential for the dissemination of lessons from Boeing's experience is much broader than from Vale and BHP's tailings dam failures, even though the organisational lessons are very similar.

There is another feature of the Boeing story that makes it of particular interest. Few industries are at risk of accidents that kill hundreds at a time, as can happen with commercial aviation. One might have thought safety would be Boeing's top priority. It often claimed it was, but this was far from the case. At Boeing, the pursuit of profit trumped safety in quite reckless ways. On the face of it this was irrational. Another sociologist, Max Weber, argued that modern capitalism and the modern corporation constituted the most rational form of economic organisation that has existed. Yet Boeing followed a course of action that called its social licence seriously into question and has enabled its arch competitor, Airbus, to draw ahead in the race to dominate the global market for civilian aircraft. How could this have happened? There are vital lessons to be learned here about how corporations operating in hazardous industries need to behave in order to ensure their long-term survival.

CHAPTER OUTLINE

There are numerous aspects to this story: engineering design, dominance of shareholder interests, human factors, regulatory failure, incentives provided to senior management, legal response, question of safety culture, and even the changing nature of capitalism. The following paragraphs provide a road map of what lies ahead. The final chapter introduces another kind of map, a causal map, that shows how these components fit together.

Chapter 2 provides a more detailed account of the two crashes, highlighting the way both aircraft behaved like bucking broncos, in slow motion. It was only after the pilots finally lost all control, that each aircraft plunged at very high speed, into the sea in one case, and into the ground in the other. The chapter identifies some of the questions to which subsequent chapters provide answers.

Chapter 3 describes Boeing's historic transition from a great engineering firm to a business run in the interests of shareholders. This was achieved, in part, by moving the corporate centre away from Seattle, where Boeing aircraft are assembled, to Chicago. The stated intention of this move was to distance top executives, both literally and figuratively, from the engineers who designed and built Boeing's aircraft. Putting it bluntly, Boeing set out to disempower its engineers. It was this move that led ultimately to the 737 MAX crashes.

Boeing's transformation is best seen as part of a more general movement from stakeholder capitalism to shareholder capitalism that originated in the Anglo-American world in the 1980s and 1990s. This period saw a fundamental transformation in the nature of capitalism, wrought by the ideology of neo-liberalism, during the terms of office of President Ronald Reagan in the U.S. and Prime Minister Margaret Thatcher in the U.K.[3]

Boeing's organisational trajectory is strikingly similar to that followed by the oil and gas giant, BP. It, too, was transformed at about the same time, from a company where engineers wielded a great deal of influence to one in which shareholder interests were paramount. And arguably, with similarly disastrous results – the 2005 Texas City refinery explosion and the 2010 Deepwater Horizon blowout in the Gulf of Mexico. Chapter 3 sketches the BP story as a way of illuminating the Boeing experience.

Chapter 4 examines just how Boeing ensured shareholder supremacy. It did so by providing long-term bonuses to its CEO and top-most executives, which depended entirely on the company's financial performance. The value of these bonuses was many times the value of the fixed salaries of these executives, ensuring that their focus on shareholder value never wavered. One of the mechanisms used to enhance shareholder value was the share buyback. This involved the company using its profits to buy its own shares on the stock market and then liquidating them, thus driving up the price of remaining shares, to the benefit of remaining shareholders and top executives. Buybacks were little more than a "license to loot". The practice meant that profits were not available for re-investment, in particular, in the development of a new aircraft to replace the 737, which was long overdue.

In the end, Boeing fell behind Airbus. When it was finally forced to act there was no time to develop a new aircraft and Boeing decided to simply upgrade the existing 737 model by adding larger, more fuel-efficient engines – the "re-engined" option. Thus was the 737 MAX born. But this modified aircraft no longer had the aerodynamic stability of the earlier version, forcing Boeing to make further modifications that led ultimately to disaster.

Chapter 5 introduces the principle of inherently safe or safer design. Where safety is not considered at the design stage, safety devices may need to be added on at a later stage, to compensate. These safety add-ons provide further opportunities for things to go wrong. In other words, these additional defences against danger may themselves be dangerous. This principle is well understood among safety professionals but was ignored by Boeing.

Scale model testing of the MAX in a wind tunnel revealed that the re-engining of the 737 had introduced a risk of stall in certain high-speed situations. Later, during test flights of the prototype, test pilots discovered a further stall risk in some low-speed situations. The solution was to add a piece of software to the flight control system that would automatically take control in these situations to avoid stall. The software would be triggered by a flimsy device – an AoA (angle-of-attack) sensor – that indicated how close the aircraft was to stalling. The device was located on the outside of the fuselage, where it was vulnerable to damage, particularly by bird strike, in which case it might become wildly inaccurate. This could cause the software to pitch the nose of the aircraft down, to counteract a perceived stall risk, when in fact there was no risk of stall at all. A defective or damaged AoA sensor played a part in both the doomed flights.

There is another important safety principle that was violated here. An aircraft should not be vulnerable to a single-point failure, such as the malfunction of a single AoA sensor. Designers should anticipate such failures and design accordingly. In other contexts, the principle is expressed slightly differently: defence against major accident hazards must be defence in depth, meaning that if one line of defence fails, a second or even third line of defence stands ready to control the hazard. Of course, Boeing engineers understood the need to avoid single-point failures that might lead to disaster, but in their rush to get the plane to market, this principle was overlooked.

Chapter 6 deals head-on with Boeing's claim that the accidents were due to pilot error and that pilots should have known better. It describes in some detail how events unfolded in the case of the Ethiopian Airlines crash. Just seconds after take-off, the aircraft apparently struck a bird, damaging the AoA sensor. This triggered a false stall warning – the joystick began to shake, which it continued to do for the rest of the flight. The pilots were preoccupied with this when, some seconds later, the defective sensor triggered a second crisis, the automatic activation of the software to push the aircraft nose downwards. The pilots were overwhelmed with a multitude of alarms in the cockpit and failed to do the one thing that might have saved them, that is, to deactivate the software that had taken control of the aircraft. It was this failure that led Boeing to assign all blame for the two accidents to the pilots. Its initial view was that the MAX was a safe aircraft if only the pilots had followed procedures.

Boeing's initial view ignored altogether the discipline of human factors, which identifies circumstances that are conducive to front-line operators making errors and ensures that designers take these into account. Boeing's more considered view recognised that the software was indeed conducive to error. It announced a series of changes to the software to protect pilots from similar experiences in future. Before the transformation of Boeing in the 1990s, the company had employed human factor specialists to deal with these issues at the design stage, but this capacity had been substantially eroded by the time the MAX was being developed.

The chapter also discusses the possibility that automation will progress to the point where pilots are no longer needed in passenger aircraft, thus solving the problem of human factors once and for all.

The MAX could only take to the skies after it had been certified as airworthy by the regulator, the FAA (Federal Aviation Administration). Chapter 7 deals with the certification process. FAA funding shortfalls meant that the FAA was often far behind in its certification work. Eventually, congress instructed the FAA to delegate aspects of the certification process to designated Boeing engineers. These FAA delegates found themselves in an acute conflict of interest, on the one hand, pressured by Boeing to speed the certification process, and on the other, expected by the FAA to represent its interests diligently. The FAA was supposed to oversee the activities of these certifiers but was unable to do so effectively. The result was that in many matters, Boeing was, in effect, self-certifying. A more clear-cut case of regulatory capture is hard to imagine.

This outsourcing led fairly directly to disaster. The software designed to deal with stall risk should never have been certified, vulnerable as it was to a single-point failure – damage to the AoA sensor. Initially, the design required a second sensor to confirm the readings of the first, before the software activated. But as time went on, new information came to light requiring that this second sensor be abandoned, leaving the software vulnerable to a single-point failure. A safety study was conducted by Boeing engineers prior to the abandonment of the second sensor. When the time came to submit a safety study to the regulator, this was the study submitted, making no reference to the subsequent version of the software reliant on a single sensor. How this could have happened is a matter of dispute. On one view, this was a deliberate act of fraud. A second view is that Boeing had played by the rules, but these rules were deficient. A third view is that what happened was the result of communication failures within Boeing, which meant that no one had the full picture and no one was truly accountable for the representations that Boeing made to the FAA about the safety of the software.

Chapter 8 explores some of the legal consequences flowing from the MAX crashes. First, the Department of Justice began proceedings against Boeing for conspiracy to defraud the FAA. It alleged a conspiracy among Boeing employees to conceal from the FAA crucial information about the stall prevention software and to ensure that there was not even a mention of the new software in manuals and pilot training material. The Department chose to hold Boeing accountable for the criminal misconduct of two particular employees, but it exonerated senior management. Boeing agreed that two of its pilots had indeed conspired to defraud the FAA, and it accepted the penalty of $2.5 billion, in return for the matter not proceeding further to a plea or conviction. It is hard to imagine these two pilots agreeing to take the rap for Boeing in this way. They were widely perceived to have been used as scapegoats.

Strangely, the Department of Justice was not prepared to let the matter rest. It compounded the perception of injustice by prosecuting one of the pilots personally for the fraud he was alleged to have committed. Because he was now the target of the prosecution, he defended himself vigorously, bringing new information to light which resulted in his acquittal by a jury. This was indeed an ironic outcome. By mounting a prosecution against this individual, and losing, the Department undermined the entire basis of its earlier agreement with Boeing. If the pilot was not guilty of fraud, neither was Boeing. Somehow, this implication appears not to matter.

A more important judgement, but one much less publicised, was handed down in a case bought by shareholders against the members of Boeing's Board of Directors. It found that before the Indonesian airline crash, the directors had utterly failed to establish a system for monitoring and overseeing airplane safety, and that after that crash, when confronted with "red flags" about the safety of the stall avoidance software, they had consciously disregarded their duty to investigate.

The Board claimed in its defence that it oversaw the quality and safety of the 737 MAX by monitoring the progress of the FAA's extensive certification review of the 737 MAX. The judge was dismissive of this claim. As he said,

> the Board focussed on the 737 MAX's production, development and certification in order to assess production timelines and revenue expectations, and to strengthen the Company's relationships with FAA officials - not to consider customer safety.

In one of his most important remarks, he goes on to say:

> the fact that the company's product [on the face of it] satisfied regulatory requirements does not mean that the board has fulfilled its oversight obligations to prevent corporate trauma.

All boards should take this comment to heart. Boards need to exercise scepticism about what they are told. They should question the good news and seek out the bad, or, as is frequently said, "challenge the green and embrace the red". This is another critically important safety lesson to emerge from many major accident inquiries, which emerges yet again from the Boeing tragedies.

As a result of these findings, Boeing entered an agreement with shareholders to make a series of organisational changes, which, if implemented in good faith, should make an enormous difference to safety. They are not spelt out here but are discussed in Chapter 8. In my view, this is the most important outcome of all the litigation following the accidents. It is to be hoped that Boeing will not lose sight of this agreement and that the various regulatory agencies will find a way to enforce its provisions.

The judgement in the shareholder case was not only about learning lessons; it was also about assigning culpability. The judge found that the Board had publicly and knowingly lied about whether and how it had monitored the safety of the MAX as it was under development. If the Department of Justice had been able and willing to prosecute Boeing's directors, the outcome would have been far more congruent than prosecuting the pilots.

Chapter 9 continues the discussion of incentives. Prior to the 737 MAX crashes and door plug blowout, safety was routinely declared to be Boeing's top priority, but there was nothing in the remuneration system to focus the attention of top executive managers on aircraft safety. It was simply taken for granted.

All this began to change, particularly after the door plug blowout. Major accidents are rare, but there are precursor events that occur relatively often. If the frequency of such events can be driven downwards, the risk of a major accident is likewise reduced. Boeing was able to identify some of the precursors to the poor-quality manufacturing that had led to the door plug blowout and to incentivise a reduction in the number of such precursor events. This now impacts the annual bonuses paid to employees at all levels. Most importantly, at the time of writing, long-term bonuses paid to the most senior executives depend to an extent on the company's efforts to control the number of precursor events. This represents a major break with the long-standing policy of tying long-term bonuses exclusively to the company's financial performance. If this new approach takes root, the travelling public will be able to breathe a little easier.

The challenge, as yet unmet, is to incentivise the quality engineering decisions that would have prevented the 737 MAX crashes.

Chapter 10 deals with the concept of safety culture. In response to the MAX crashes, Congress passed an Act in 2020 which among other things effectively required Boeing, and companies like it, to implement a positive safety culture. This is a controversial concept that has led to endless debates in the safety literature. It can, however, be boiled down to a single idea – a good reporting culture. That in turn implies many of the other elements that have been taken to be part of a good safety culture, for example, that it must be a *just* culture. In relation to reporting, it must be a *no-blame* culture.

A Congressional Expert Panel set up to investigate these matters found that Boeing had failed to create a culture of reporting in part because it had not convinced employees that it was safe to make reports.

Chapter 10 also lays out, very briefly, the principles of an effective reporting system,[4] drawn from the literature on high-reliability organisations, which in many ways parallels the safety culture literature.

Finally, to return to the 2020 Act, perhaps its most important provision is a requirement that Boeing and similar companies set up a comprehensive safety management system. What Congress failed to recognise was that such a system requires a strong regulatory oversight regime if it is to be effective. Otherwise, it runs the risk of being no more than a paper tiger.

Finally, Chapter 11 provides a causal diagram showing how all the various contributory factors interacted to generate disastrous outcomes. It may be useful to refer to this from time to time while reading. Chapter 11 highlights one factor as having been particularly influential – "the priority of shareholder returns over safety" – and it argues that unless this can be turned around, passengers will remain at risk. It recommends the use of criminal sanctions for the failure of CEOs and board members to exercise reasonable care, as a way of counterbalancing the enormous pressures on these people to maximise shareholder returns.

NOTES

1 Recounted in Hopkins A and Kemp D, *Credibility Crisis: Brumadinho and the Politics of Mining Industry Reform*, Wolters Kluwer, Sydney, 2021.
2 Toby J, "Is Punishment Necessary", *Journal of Criminal Law and Criminology and Police Science*, 55, p. 332. (1964).
3 This transition has been admirably described, most recently, by Varoufakis Y, *Technofeudalism – What Killed Capitalism. Bodley Head*, London, 2024.
4 See Hopkins A, "A Practical Guide to Becoming a High Reliability Organisation" http://hdl.handle.net/1885/287366

2 The Crashes

The Lion Air 737 MAX had just taken off from Jakarta International airport on the morning October 29, 2018, when the pilot's control column (joystick) began to shake. This was an indication – false in this case – that the aircraft was about to stall. Then, the aircraft's nose mysteriously dropped. The captain squeezed a switch on the control column to push the nose back up. It lifted, but then dropped again. This sequence was repeated again and again –21 times in all. The aircraft seemed to have a mind of its own and appeared intent on diving into the sea. It must have been like riding a bucking bronco, in slow motion. The terror of the passengers is hardly imaginable.

The pilots had no idea what was happening. It turned out that a piece of automated software had activated, because of an equipment malfunction. The software had taken charge of the flight control system and repeatedly and forcefully angled the nose downwards. The battle for control of the aircraft ended after eight minutes when the plane plunged almost vertically into the sea at very high speed. All 189 people on board perished.

The equipment that malfunctioned was a tiny sensor on the side of the aircraft that was giving false readings. It was designed to measure the angle of attack – the angle of the wing against the oncoming airstream. If the angle was too great, this meant the aircraft was climbing too steeply and was in danger of stalling; if the angle was too low it meant that the aircraft might be losing height. The false readings from the sensor were transmitted to the flight control system which automatically took over from the pilots.

As it happened, the sensor had malfunctioned on the *previous* flight but, by a stroke of good luck, the pilots had managed to retrieve the situation.[1] The captain of that flight reported the problem, as the pilots had experienced it in the cockpit, namely, that the aircraft nose had suddenly and inexplicably pitched downwards. They did not know that this stemmed from the faulty sensor. A maintenance crew on the ground in Jakarta was also unable to pinpoint the cause of the problem and cleared the aircraft to continue with its next sector. Neither they nor the captain knew that *prior* to the previous sector, a maintenance team in Bali had replaced the sensor but failed to check that it was functioning properly.[2] It wasn't.

The 737 MAX was a brand-new aircraft, a substantially modified version of the 737 that had been flying since 1967. Boeing had great faith in this new version, which it called the 737 MAX. In the weeks that followed, it refused to acknowledge that there was anything wrong with the aircraft or its flight manuals. According to the CEO, "The bottom line here is that the 737 MAX is safe. Safety is a core value for us at Boeing".[3] He said the company had provided the pilots "with all the information that's needed to fly our airplanes".[4] In so saying, he was implicitly blaming the pilots for the crash. Boeing also highlighted the mistakes by the Lion Air maintenance crews as the beginning of the chain of errors.[5] It is true that these teams made mistakes, but highlighting these mistakes obscures far more than it reveals. It was a

DOI: 10.1201/9781003629269-2

further attempt to shift any blame away from Boeing and onto the employees of a foreign airline. This blame-shifting exercise was successful, at least in the short term. The CEO's statements convinced investors that the aircraft was fundamentally safe, and within days Boeing's share price had risen to above its pre-crash level![6]

Investigations continued behind the scenes, but the Lion Air crash soon disappeared from the headlines. After all, the crash involved a foreign airline, with foreign pilots and no Americans killed. Questions about the safety of the aircraft had apparently been settled. Indeed, a representative of the U.S. regulator felt able to say that the crash was a "one-off", caused by poor pilot performance.[7] Boeing continued to exude confidence in its aircraft, and its CEO announced he was hoping to increase the assembly line production rate from 52 to 57 planes a month.

But pilots, and in particular U.S. pilots, did not accept that poor pilot performance was the explanation. In fact, they were furious. None of them had been trained on the new automated flight software. They did not even know it had been fitted to the MAX aircraft!

Then, just over four months after the Lion Air crash, it happened again. On Sunday, March 10, 2019, an Ethiopian Airlines 737 MAX crashed immediately after take-off from Addis Ababa airport, killing all 157 on board. The plane had run into the exact same problem as the Lion Air plane. Aware of the Lion Air experience, the pilots had guessed immediately that the problem was a malfunctioning sensor. But they were nevertheless unable to gain control of their aircraft, which crashed at high speed into the ground. Boeing again went into denial. Within a day it issued a statement that "the 737 MAX is a safe airplane that was designed built and supported by our skilled employees who approach their work with the utmost integrity".[8] The U.S. regulator, the FAA (Federal Aviation Administration), was not immediately willing to ground the aircraft, even as a precautionary measure, but regulators around the world had had enough. The Chinese grounded the MAX within 24 hours.[9] A day later, regulators in the European Union, India, Australia, Singapore, and Canada followed suit. On the following day, the FAA was provided with evidence that the behaviour of the Ethiopian aircraft immediately before the crash was very similar to the behaviour of the Lion Air aircraft. This pointed undeniably to the same problem – a malfunction of the angle of attack sensor. This was finally enough to convince the FAA, and it grounded the 737 MAX in the U.S. forthwith, three days after the crash. The grounding remained in place around the world for nearly two years. The inadequacy of regulatory oversight and its contribution to these accidents is a matter to be taken up later.

This account has already begun to raise questions about Boeing's own contribution to the crashes. Why had the pilots not been trained on the new software? And why was the aircraft so vulnerable to the failure of a single small piece of equipment – an airflow sensor? The reader will have already intuited that this was a fundamentally unsafe design. Aircraft designers would tell you that a design that allowed a single-point failure to bring down an aircraft is unacceptable. Yet Boeing had accepted this situation.[10] Why? I shall return to these questions in due course.

Finally, we fast forward nearly five years, to January 5, 2024. The 737 MAX fleet is flying again. An Alaskan Airlines 737 MAX had just taken off from an airport in the state of Oregon bound for an airport in Southern California. Suddenly, a whole side panel blew out, leaving a door-sized hole in the side of the aircraft. The air rushed out through the hole and would probably have sucked out any unbelted passengers sitting nearby. Fortunately, the closest two seats were empty. The aircraft made an emergency descent and landing with no one injured. However, the potential for multiple fatalities was obvious. It was quickly discovered that four bolts that were supposed to hold the panel in place had either been loose or missing altogether. Various airlines immediately cancelled flights until they had had a chance to inspect their own fleets, and the FAA followed suit the next day grounding all aircraft of the specific type – 737 MAX 9 – until inspections could be carried out. United Airlines announced it had found loose bolts on some of its aircraft.[11] It later turned out that all four bolts were missing on the Alaskan Airlines plane.

This was a very different type of safety issue from that which had brought down the Indonesian and Ethiopian airliners. Those tragedies involved a new aircraft design which had not been adequately explained to pilots and was in any case not inherently safe.[12] The problem of loose or absent bolts was something that had originated on the assembly line. How had this happened and how is it that Boeing's quality control procedures had failed to pick up that this was happening? Again, these questions will be answered in due course. We shall see that they stem from the same root causes.

NOTES

1 Robison, pp. 272–273. My account at this point relies heavily on the detailed analysis in the book by Robison P, *Flying Blind*, Penguin Random House, New York, 2021.
2 Robison P, *Flying Blind*, Penguin Random House, New York, 2021, p. 272.
3 Robison P, *Flying Blind*, Penguin Random House, New York, 2021, p. 281.
4 Robison P, *Flying Blind*, Penguin Random House, New York, 2021, p. 282.
5 Robison P, *Flying Blind*, Penguin Random House, New York, 2021, p. 297.
6 Robison P, *Flying Blind*, Penguin Random House, New York, 2021, p. 297.
7 Robison P, *Flying Blind*, Penguin Random House, New York, 2021, p. 304.
8 Robison P, *Flying Blind*, Penguin Random House, New York, 2021, p. 312.
9 Robison P, *Flying Blind*, Penguin Random House, New York, 2021, p. 313.
10 Robison P, *Flying Blind*, Penguin Random House, New York, 2021, p. 323.
11 "Alaska Airlines Timeline: How Incidence Involving Missing Door Plug Unfolded." ABC News, January 10, 2024. https://abcnews.go.com/US/timeline-incident-involving-missing-door-plug-boeing-737/story?id=106191494
12 See Chapter 5.

3 Boeing's Historic Transition

The start of Boeing's troubles can be traced to organisational changes made in the late 1990s. These changes reflected an ideological tidal wave that swept the Western world from the 1980s onwards – the tidal wave of neo-liberalism – to be explained shortly. The story has been told many times but needs to be summarised here.

Boeing was founded in 1916. It was led by engineers whose focus was on engineering excellence. It was listed on the NY Stock Exchange in 1962, but it retained a focus on engineering quality above all else. As one observer noted:

> For about 80 years, Boeing basically functioned as an association of engineers. Its executives held patents, designed wings, spoke the language of engineering and safety as a mother tongue. Finance wasn't a primary language. Even Boeing's bean counters didn't act the part. As late as the mid-'90s, the company's chief financial officer had minimal contact with Wall Street.[1]

All this began to change in 1997 when Boeing bought out a competitor, McDonnell Douglas, which at that stage was on the verge of failure. Ironically, this acquisition amounted to a reverse takeover, since McDonnell Douglas senior executives ended up on top. Boeing was now Boeing in name only. McDonnell Douglas had been focussed on commercial success above all else and the new leadership quickly began changing Boeing's priorities.

But Boeing's Seattle-based engineers did not give up without a fight. In order to finally win the battle, the new leadership shifted Boeing's headquarters to Chicago in 2001. As the CEO at the time explained:

> When the headquarters is located in proximity to a principal business—as ours was in Seattle—the corporate centre is inevitably drawn into day-to-day business operations.

The move to Chicago was designed to put an end to this possibility. It separated engineers, whose concern had always been quality, from senior managers, whose prime focus was commercial success. After the move, engineers feared raising their concerns, lest they be disciplined in some way.[2]

The fact is the move was deliberately designed to disempower these engineers and ensure that shareholder interests were paramount. Soon after, a new CEO was able to say:

> When people say I changed the culture of Boeing, that was the intent, so that it's run like a business rather than a great engineering firm.[3]

But having senior executives involved in day-to-day manufacturing operations, as they had been in Seattle, had benefits that the move to Chicago overlooked. Previous Boeing leaders had prided themselves in knowing everything about the aircraft they

DOI: 10.1201/9781003629269-3

built. One made a point of leaving the factory through a different door every night so that he could talk to different people on the production line. He invited mechanics to have lunch with him once a week without their bosses present, so that they might speak their minds more freely about what might be going wrong.[4] Another senior executive is reported to have said, "the only thing that will make me rip off your head and shit down your neck is withholding information". Burying things and letting them fester was just not acceptable.[5] Boeing had even set up systems to ferret out the bad news as early as possible, so as to avoid confusion and costly changes.[6]

These attitudes are the basis of much present-day thinking about safety in hazardous industries. In particular, the need for senior executives to talk to front-line workers is routinely recommended. One of the most influential statements of this philosophy was provided by Lord Cullin in his inquiry report into a rail crash at Ladbroke Grove in the U.K. in 1999 that killed 31 people and injured a further 417. He wrote

> Companies in the rail industry should be expected to demonstrate that they have, and implement, a system to ensure that senior management spend an adequate time devoted to safety issues, with front line workers. Companies should make their own judgement on how much time their leaders should spend in the field, but best practice suggests at least one hour per week should be formally scheduled in the diaries of senior executives for this task. Middle ranking managers should have one hour per day devoted to it, and first line managers should spend at least 30% of their time in the field.[7]

Very few senior executives demonstrate best practice in this respect. And certainly, it would be virtually impossible for Boeing's leaders in Chicago to talk to Boeing workers in this way. Yet in days gone by, at least one of Boeing's leaders did so on a daily basis. It is hard to imagine a more profound transformation in the behaviour of senior executives.

Interestingly, in mid-2024, a new CEO showed signs of wanting to return to some of Boeing's older ways.[8] He chose to set up his home in Seattle and to work out of Boeing's offices there. An engineer by training, he had a reputation for walking the shop floor and communicating well with employees about their issues. He was also reputed, in a previous job, to have abolished privileged parking for senior executives, requiring them to take their chances in the general employee parking lot. This was perhaps only symbolic recognition of the need to bring senior executives and employees closer together. However, it led one employee to say, "it was a little thing but had a big impact on me".

In addition to disempowering its engineers by moving head office to Chicago in 2001, Boeing hived off parts of its manufacturing operations in 2005 into a separate listed company, Spirit AeroSystems. This meant that manufacturing previously done in-house was now outsourced to an independent company. This in turn meant that Boeing was able to drive hard bargains with Spirit and so reduce its own costs. It also meant that Boeing was later able to point the finger at Spirit for various manufacturing defects, which were the almost inevitable consequence of the financial pressure it had placed Spirit under. Under this new arrangement, Boeing's role was reduced to assembling components acquired from Spirit and other suppliers. When Boeing eventually set about repairing the damage done by this transformation, it reacquired most of Spirit AeroSystems.[9]

But to return to the main line of argument here: it was only after the two crashes that Boeing realised it had gone too far in disempowering its engineers. It decided to re-organise its reporting lines, so that engineers no longer reported to commercial managers. Instead, they would report to more senior engineers, and ultimately to a Chief Engineer who reported directly to the CEO.[10] As explained by one of the lawyers involved in these events,

> The value of that improvement is that folks that have the technical knowledge and can spot issues have a centralized way they can report them, so that they are insulated from interference by business leaders who may be more swayed by economic or bottom-line considerations.[11]

This was a radical change in the strategy that Boeing had been following since the move to Chicago – a complete about-face. It was a recognition that the organisational structure it had created had in turn contributed to the two crashes.

However, despite this recognition of what needed to be done, the restructuring had not been effectively achieved when an external review panel published its report in February 2024.[12] This inevitably undermines Boeing's credibility on the issue.

A STRIKING PARALLEL

At this stage, it is helpful to set Boeing in a broader context. As it happens, Boeing's organisational trajectory, including its about-face after the 737 crashes, is strikingly similar to the path followed at about the same time by the U.K.-based oil and gas company, BP. The full significance of the Boeing story is best grasped by identifying the similarities in the paths followed by BP and Boeing.[13]

Until the beginning of the 1990s, BP had been a technically focussed and highly centralised company. Its assets (oil fields, refineries, etc.) around the globe were required to conform to a set of global standards, and various technical operations, such as seismic surveys and well-drilling operations, were carried out by central BP divisions. In 1989, an up-and-coming executive, John Browne, took charge of BP's exploration and production worldwide.[14] (BP's other major activity, refining and marketing, was not yet under his control.) Browne was determined to convert BP from a technically run company to a commercial one. Within a year, he began a process of *decentralisation* in which BP's major oil fields became business units, in effect independent companies, responsible for their own destiny. They would be encouraged to look outside BP for the provision of technical services, from whomever could provide them most competitively. In short, the provision of such services would be outsourced, and BP's central technical divisions would be largely done away with.[15] The "primary role [of this streamlined BP] would be to decide which projects to back and where possible have others do the actual work".[16] As a consequence, engineers wielded far less influence than previously and engineering standards slipped, paving the way for the disasters that subsequently befell BP. One commentator goes further: "the old values like loyalty and paternalism were jettisoned and a harsher atmosphere descended on the company".[17]

In 1995, Browne took over as CEO of the whole company and immediately extended his model to refining and marketing. Big refineries were now treated as independent business units and could source the oil they refined from wherever they wanted.

The result of this restructuring was a massive improvement in financial performance. Investors and business analysts were delighted. Management experts from U.S. business schools came to study BP's methods and wrote glowingly about them. Browne was voted the U.K.'s most successful businessman.

CAPITALISM IN TRANSITION

It is no coincidence that Boeing and BP pursued the policies of commercialisation at about the same time. In the 1980s and 1990s, capitalism in the Western world was in transition. For present purposes, it is useful to identify two types of capitalism: stakeholder and shareholder capitalism. Where companies operate in the interests of all interested parties – shareholders, workers, customers, vendors, governments, and others, we can speak of stakeholder capitalism. Where they operate in the interests of shareholders alone, we can speak of shareholder capitalism. Under shareholder capitalism, companies may take account of the interests of other groups, but only so far as it is in the interests of shareholders to do so. It is sometimes noted that the longer the time interval that is considered, the more these models converge, since in the long run companies will only thrive if they treat all their stakeholders fairly and with respect.[18] But there is no doubt that in the short term, shareholders may derive the greatest financial benefits from short-sighted and less than scrupulous corporate behaviour.

The 1980s and 1990s saw a movement from stakeholder to shareholder capitalism. This was not dictated by changes in the law,[19] but rather by an ideology, or philosophy, sweeping the Anglo-American world – neo-liberalism[20] – known in the U.K. as Thatcherism and in the U.S. as Reaganomics. In the world of corporate governance, neo-liberalism was associated with the name of Milton Friedman. Here is how one writer describes Friedman's contribution.[21]

> Shareholder primacy, or the idea that a corporation is only responsible for increasing shareholder value, was made popular by Nobel prize-winning economist Milton Friedman in the 1970s. He argued that executives work for the owners (shareholders) and *the only social responsibility of a business is "to use its resources and engage in activities designed to increase its profits,* so long as it stays within the rules of the game, which is to say, engages in open and free competition without deception or fraud." (emphasis added).

Friedman dismissed anything else as "pure and unadulterated socialism".[22]

Friedman's notion of the social responsibility of a business is quite the opposite of present-day ideas of "corporate social responsibility", but it resonated with business leaders in the 1980s and 1990s.[23] The transformations of Boeing and of BP in the 1990s are clear expressions of this dominant ideology.[24]

THE IMPACT OF SHAREHOLDER CAPITALISM ON SAFETY

Returning to BP, one aspect of its transformation that went largely unnoticed at the time was that there was no longer centralised responsibility for the management of health, safety, and the environment. These matters were devolved to oil field and refinery managers. The problem with this approach is that catastrophic events such as refinery explosions or oil well blowouts – as well as aircraft crashes – are relatively rare and there was very little way of knowing how well these risks were being managed. The result was that "BP had become a carefully constructed time bomb, whose ticking was drowned out by the roar of tributes to Browne."[25]

Perhaps a minefield might be a better metaphor since BP began to experience a series of calamitous events. The two best known were the explosion and fire at the Texas City Refinery in 2005 that killed 15 people, and the blowout in the Gulf of Mexico in 2010 that killed 11 men and did massive environmental damage around the coast of the Gulf of Mexico – the infamous Deepwater Horizon blowout. This latter event eventually cost BP at least $65 billion dollars, threatening the company's very existence.[26]

The disempowerment of engineers brought about by BP's organisational transformation was a key contributory factor to both these disasters.[27] BP finally understood this after the Deepwater Horizon disaster. Accordingly, it re-organised its engineers into separate reporting lines, independent of the business units in which they worked, and it created an entirely new Safety and Operational Risk function, answerable directly to the CEO and charged with the responsibility of ensuring that major accident risk was being properly managed within the business units.[28] To this end, many of the staff employed in this new function were "embedded" in the business units, that is, physically located in these units and in daily contact with their operations, although remaining answerable up the Safety and Operational Risk line.

Boeing's announced recentralisation of its engineering function after the 737 MAX crash is entirely analogous to BP's recentralisation of its technical staff ten years earlier. If only Boeing had been able to learn from BP's experience, 346 people might not have lost their lives. There is a challenge here for business schools that lionised BP's and Boeing's commercial successes prior to their disastrous accidents. They should also be teaching their students about how companies can best organise themselves to avoid catastrophic accidents. To the extent that business schools fail to do so, they are missing a critical opportunity to contribute to the prevention of such accidents.

NOTES

1 Useem J, "The Long-Forgotten Flight that Sent Boeing Off Course", *The Atlantic*, February 13, 2020.
2 Useem J, "The Long-Forgotten Flight that Sent Boeing Off Course", *The Atlantic*, February 13, 2020.
3 Robison P, *Flying Blind*, Penguin Random House, New York, 2021, p. 135.
4 Robison P, *Flying Blind*, Penguin Random House, New York, 2021, p. 122.

5 Robison P, *Flying Blind*, Penguin Random House, New York, 2021, p. 70.
6 Robison P, *Flying Blind*, Penguin Random House, New York, 2021, p. 68.
7 Cullen L., *Ladbroke Grove Rail Inquiry*, part 2, 2001, p. 64.
8 "Boeing's new CEO inherits a big mess", *The Wall Street Journal*, August 10–11, 2024.
9 "Boeing to acquire troubled supplier Spirit AeroSystems, with Airbus parts split off", *Seattle Times*, June 30, 2024.
10 See Chapter 8.
11 Tokar, "Prosecutors Credited Boeing for Compliance, Organizational Reforms", *The Wall Street Journal*, January 15, 2021.
12 "Section 103 Organization Designation Authorizations (ODA) for Transport Airplanes", Expert Panel Review Report, FAA, February 26, 2024, p. 37.
13 This account draws heavily on Bergin, T, *Spill and Spin: The Inside Story of BP*. Random House, London, 2011.
14 Bergin, T, *Spill and Spin: The Inside Story of BP*, p. 33.
15 Bergin, T, *Spill and Spin: The Inside Story of BP*, p. 22.
16 Bergin, T, *Spill and Spin: The Inside Story of BP*, p. 25.
17 Bergin, T, *Spill and Spin: The Inside Story of BP*, p. 26.
18 Australian Institute of Company Directors, General Duties of Directors.
19 Clarke T, "The Contest on Corporate Purpose: Why Lynn Stout Was Right and Milton Friedman Was Wrong", p.17/49.
20 "Neo" in contrast to the philosophy of liberalism that emerged towards the end of the 17th Century.
21 D'souza D, "What Is Stakeholder Capitalism?", October 3, 2022, https://www.investopedia.com/stakeholder-capitalism-4774323
22 Lazonick W and Sakinc ME, "Make Passengers Safer? Boeing Just Made Shareholders Richer", Prospect Magazine, May 31, 2019. https://prospect.org/environment/make-passengers-safer-boeing-just-made-shareholders-richer./
23 See the novel by Wolfe T, *The Bonfire of the Vanities*,Vintage, London,1987, for a horrifying description of the greed of the 1980s.
24 One of the stakeholder groups that lost out most conspicuously in Boeing's transition to shareholder capitalism was its workforce. Prior to Boeing's change of course, its workers were well off, but in recent decades their pay and conditions have been steadily whittled away by cost cutting, and Boeing jobs are no longer coveted in the Seattle area as they once were. See "Generations of Workers Coveted Boeing Jobs. Strike Reveals How Much Has Changed", *Washington Post*, September 17, 2024. https://www.washingtonpost.com/business/2024/09/17/boeing-strike-union-workers-contract/
25 Bergin, T, *Spill and Spin: The Inside Story of BP*, p. 38.
26 "BP's Deepwater Horizon Costs Reach $65 Billion", The Maritime Executive, January 16, 2018, https://www.maritime-executive.com/article/bp-s-deepwater-horizon-costs-reach-65-billion
27 I shall not discuss this here, but for a detailed account, see my book, *Disastrous Decisions*.
28 See my book, *Organising for Safety: How Structure Creates Culture*, Chapter 5.

4 Shareholder Supremacy and the 737 MAX

The strategy for assuring shareholder supremacy under shareholder capitalism is to align the interests of the CEO and other senior executives with those of shareholders. This is done by providing top executives with a variety of incentives tied to the long-term financial performance of the company – so-called *long-term incentives*. It is important to know how these operate because they are the key to understanding the decision-making that led to the development of the 737 MAX aircraft.

Boeing uses several mechanisms to achieve this long-term incentivisation, all of which are commonly used by large publicly listed companies. I shall mention three.

The first and easiest to understand is to reward executives with bonus shares that are withheld for three years after their award. If the company prospers in that time, the shares are worth correspondingly more when they are finally transferred. They can then be sold at a profit. In the case of Boeing's CEO, he is not permitted to sell these shares until he leaves the company, linking his financial interests even more firmly to those of the company.[1]

A second mechanism is to provide a bonus in the form of stock options. These are options, granted at a certain point in time, to buy company shares, at a later time, at roughly the same price they were when they were granted. If the share price has gone up in the intervening time, the recipient can exercise the option to buy at the lower price and then resell, making a profit. If the share price has stagnated or retreated, the options are worthless. Again, therefore, it is in the interest of the recipient that the share price rise.[2]

A third mechanism that Boeing has used encourages the company to engage in a virtual, but cut-throat competition with a peer group of companies, in terms of total returns they provide to their shareholders. Simply put, total returns mean dividends plus any increase in share value. In 2022, Boeing selected a peer group of 19 U.S.-based companies of comparable size and operating in similar industries. One European company was added to the list, because it is the nearest thing Boeing has to a direct competitor – Airbus. These companies were ranked in terms of their total shareholder returns. This ranking was then used to determine the value of certain of the components of the long-term bonus program, as follows. What was at stake was a "target bonus". It is not entirely clear from the documents how this target was derived, but for present purposes, this is not important. What is important is that if Boeing ended up in the bottom quartile of the ranking (bottom quarter of the peer group), the payout would be zero; if it reached the 25th percentile, the payout would be 25% of the "target" bonus; if it reached the 55th percentile, the full "target" bonus would be paid; and reaching the 95th percentile or above was treated as maximum performance and the payout would be double the "target" bonus.[3] This strategy meant that it was not enough to provide investors with a healthy total return on their

DOI: 10.1201/9781003629269-4

investments. What was important was the *relative* total shareholder return, that is, relative to the peer group of companies. This created a virtual competition between Boeing and other members of the peer group, which increased the incentive for senior executives to achieve the highest possible returns for investors. Unfortunately for Boeing's CEO and fellow senior executives, in 2022 the company was in the bottom quartile – 17th of the 20 in the comparison group. This was the third year in a row that the company had failed in this way. This particular bonus mechanism was abandoned soon afterwards,[4] but it is symptomatic of the way boards, under the influence of remuneration consultants, seek to increase pressure on their senior executives to act in the financial interests of shareholders.

As I shall explain in more detail in Chapter 9, the value of long-term bonuses for top executives greatly outweighed any other form of remuneration. They were designed to ensure that the aim of senior executives was to maximise returns to shareholders, above all else.

One way to increase returns to shareholders is to increase company profits, for example, by selling more or better products.[5] But there is another mechanism that has nothing to do with the intrinsic value of the company or what it produces – the share buyback. This involves a company buying shares in itself and then liquidating them.[6] This is conceptually paradoxical – what does it mean to buy a share of oneself? In fact, it is almost a conjuring trick. Buying and then liquidating shares result in fewer shares in the market. Assuming no change in the intrinsic value of the company, this means each remaining share will be worth more, benefiting all remaining shareholders. This will be popular with shareholders and will also result in increased bonuses for the senior executives. However, if done on a large enough scale, it amounts to market manipulation. For this reason, it was largely illegal in the U.S. until 1982. In that year, the second of Ronald Regan's presidency, the Securities and Exchange Commission (SEC) promulgated a new rule that greatly facilitated share buybacks. According to one commentator, the new rule was simply a "license to loot".[7] The problem is that the more a company uses its profits to increase the return to shareholders, the less it has to invest in future ventures. This can be disastrous for a manufacturing company like Boeing.

This shift to shareholder capitalism had damaging consequences for Boeing. By 2002, it had become clear that the company needed a radically new aircraft design to replace the 737.[8] The 737 was no longer the premier product of its type; the European Airbus A320 was.[9] Boeing could not afford to rest on its laurels. It began investigating options for a totally new "clean-sheet" replacement. However, it chose to defer any decision on this indefinitely.

At around this time, Boeing was heavily engaged in a stock buyback program. Between 2004 and 2008, when the Global Financial Crisis intervened, Boeing spent $11 billion in buybacks. During the same period, it paid out $4.8 billion in dividends.[10] This was a bonanza for shareholders, including, of course, Boeing's senior executives.

According to one commentator, "If Boeing had funded a well planned and executed investment in a wholly new aircraft, the company could have generated significant revenues and profits far into the future".[11] But a wholly new aircraft might have cost $20 billion,[12] which would have impacted Boeing's buyback program. So, the clean-sheet replacement was not actively pursued.

A brand-new replacement for the 737 remained under consideration, however, and in February 2011, Boeing's CEO was able to say:

> We're gonna do a new airplane. We're not done evaluating this whole situation yet, but our current bias is not to re-engine, it is to move to an all-new airplane at the end of the decade [ie, 2020], or the beginning of the next decade.[13]

"Re-engining" was a reference to another possible option, namely, to modify the existing 737 by equipping it with larger and more fuel-efficient engines. In December 2010, Airbus had begun offering airlines a new design called the A320neo, for "new engine option". The new engine would be larger and more fuel efficient. Boeing was at first sceptical about the A320neo, believing that its own customer base would wait until the clean-sheet replacement for the 737 was available.[14] But by mid-2011, Airbus had won more than 1,000 orders from airlines around the world. Worse than this, American Airlines, which had always bought Boeing aircraft exclusively, told Boeing it was proposing to buy hundreds of the new Airbus A320s. This caught Boeing completely off guard. It was a bombshell. Boeing panicked. Within a week, it made a counterproposal for a hypothetical, re-engined 737.[15] This was a desperate catch-up move, not properly thought through. But investors were happy: the new "program would cost $2.5 billion compared with the $20 billion for a full-blown replacement".[16] As well as being cheaper, a re-engined 737 could be delivered much more quickly than a brand-new aircraft.

American Airlines responded to this new offer by splitting its order between the A320neo and the modified 737, soon to be christened the 737 MAX. But to Boeing's great disappointment, most of American's new fleet would be Airbus.

That is how the 737 MAX was born. The delay in developing a replacement aircraft, which Boeing's share buyback program had made almost inevitable, meant that when the crisis point arrived, it was too late. Boeing completely reversed the announcement made by the CEO a few months earlier. Moreover, part of Boeing's sales pitch for its modified 737, indeed its promise to airlines, was that pilots would be able to transition to it without undergoing any additional simulator training. Simulator training is expensive and would have involved appreciable additional expense for the airlines. This was therefore an attractive feature of the re-engined 737 for Boeing's customers. This promise would later prove to be Boeing's undoing.

The prospect of a re-engined 737 proved to be very popular with the airlines. "By the end of 2012, Boeing had booked more than 2,500 MAX orders, worth an estimated $140 billion in future revenues. With years of cash flow apparently assured, in 2013 Boeing's top management embarked on massive distributions to shareholders through dividends and, in far larger sums, stock buybacks". The result was dramatic. "Boeing's share price increased by a multiple of 6.7 between January 1, 2013 and March 1, 2019, when, just ten days before the second crash, it hit a record high". Boeing had become a "true dividend rockstar".[17]

But the hasty decision to produce a re-engined 737 initiated a series of events culminating in the two MAX crashes. Subsequent chapters will detail this pathway to disaster.

THE EARNINGS CONFERENCE CALL

I cannot conclude this chapter without referring to some of the more subtle but very powerful social pressures that can magnify the financial pressure under which CEOs operate. The following paragraphs provide a close-up illustration.

Every three months, Boeing holds an online meeting, called an earnings conference call, at which the CEO and CFO (Chief Financial Officer) address investor representatives and financial analysts on the financial state of the company. The proceedings are broadcast, so this is a very public event. The call for the fourth quarter of 2023 enables us to see this social pressure in action. The call took place in late January 2024, just weeks after the door plug blowout on the Alaskan Airlines flight. Following that incident, the Federal Aviation Administration (FAA) grounded the 737 MAX 9[18] fleet for weeks. As already noted, the issue was inadequate quality control on the production line. The grounding had just been rescinded at the time of the call.

The CEO began his remarks by saying that his focus in the meeting would be on this quality issue and that the CFO would then talk about the financials. He mentioned what the company was doing to improve quality control on the assembly line and noted that the FAA had capped the production rate of 737s at 38 per month and that this would remain in force until the FAA, and Boeing, were satisfied with the quality of the manufacturing process. Importantly, he said that Boeing was not issuing a financial outlook for 2024 at the time of the call since that depended on when they could speed up the production line and "we won't predict timing. We won't get ahead of our regulator".

Despite this explanation, questions from the audience were all about the figure of 38 per month: when did he expect they would begin increasing it?; what were the prospects for ramping up production to 50?; what were the implications for the "profit profile of the MAX line"?; and so on. We can see here the pressure that the company was under from investors and financial analysts. The CEO was able to resist this pressure because the slowdown was ordered by the FAA. But it is a fair bet that in the absence of this FAA order, Boeing would have been trying to run the assembly line at a faster rate, even though the quality issues had not been fully resolved. Indeed, the CEO said as much:

> I'm glad [the FAA] called out a pause, because that is a good excuse to just take our time, do it right, and I wish I had called that out on the first day...[19]

This response shows just how vital it is that a regulator be prepared to use its powers to stop the job, or at least slow the job, in the interests of safety. And almost paradoxically, the CEO acknowledges the importance of the regulator's requirement in enabling him to overcome investor pressure.

It is truly disturbing to know that despite all the efforts made since the door plug blowout to prioritise quality over quantity, the investor pressure to maximise production has remained as strong as ever. More recent reports from Boeing's assembly lines indicate that there is still pressure to keep the production line moving at all costs.[20]

NOTES

1 Boeing 2023 Proxy Statement, p. 42.
2 Boeing 2023 Proxy Statement, p. 42.
3 A sliding scale operated between these fixed points.
4 Boeing 2023, Proxy Statement, p. 40.
5 Lazonick W, *Investing in Innovation*, Cambridge Elements: Corporate Governance, CUP, 2023.
6 Also described as cancelling or retiring them.
7 Lazonick W and Sakinc M, "Make Passengers Safer? Boeing Just Made Shareholders Richer", *The American Prospect Magazine*, May 31, 2019.
8 Robison P, *Flying Blind*, Penguin Random House, New York, 2021, p. 168.
9 Robison P, *Flying Blind*, Penguin Random House, New York, 2021, p. 166.
10 Lazonick W and Sakinc M, "Make Passengers Safer? Boeing Just Made Shareholders Richer", *The American Prospect Magazine*, May 31, 2019, p. 7.
11 Lazonick W and Sakinc M, "Make Passengers Safer? Boeing Just Made Shareholders Richer", *The American Prospect Magazine*, May 31, 2019.
12 Robison P, *Flying Blind*, Penguin Random House, New York, 2021, p. 174.
13 Ostrower J, "Boeing Boss Green-Lights All-New Next Generation Narrowbody", *Flight Global*, February 10, 2011.
14 Ostrower J, "Boeing Boss Green-Lights All-New Next Generation Narrowbody", *Flight Global*, February 10, 2011.
15 Robison P, *Flying Blind*, Penguin Random House, New York, 2021, p. 171.
16 Robison P, *Flying Blind*, Penguin Random House, New York, 2021, p. 174.
17 Lazonick W and Sakinc M, "Make Passengers Safer? Boeing Just Made Shareholders Richer", *The American Prospect Magazine*, May 31, 2019, pp. 2–3.
18 This is a variant of the 737 MAX that was grounded in 2019.
19 The Boeing Co Q4 2023 Earnings Call, January 31, 2024.
20 "On Boeing's Factory Floor, Workers Feel 'Overmanaged and under-Supported'", *The Seattle Times*, September 1, 2024.

5 Inherent Safety and the Dangers of Safety Add-Ons

Fitting larger engines to the existing 737 design introduced safety risks that would have been avoided by designing a new aircraft from scratch. It was Boeing's failure to recognise and respond adequately to these risks that brought down two new 737 MAXs in quick succession. This chapter will demonstrate these claims.

This chapter also illustrates some vital safety principles, well understood by safety professionals, but too often ignored where commercial or other interests prevail. These principles include *inherent safety* and the *dangers of safety add-ons*. These two ideas are different ways of saying much the same thing. Boeing's experience is a powerful demonstration of their importance.

The terms "inherent safety" or "inherently safer" need clarification. They are relevant at the *design* stage of processes and equipment. Hence, the discussion is usually about the "inherently safer *design*". If safety is considered only after a design has been settled on, it will usually depend on add-ons. A simple example will be useful here. If the risk of a house burning down during a wildfire is considered only after the house is built, fire safety will depend on add-ons, such as sprinkler systems, water tanks in case the town water supply fails during the fire, electrical water pumps, and backup internal combustion pumps in case the electricity supply fails. But if the risk of fire is considered prior to construction, the risk can be largely designed out by the use of non-flammable materials.

The idea of inherently safer design was brought to prominence in the chemical and process industries after a chemical plant at Flixborough, U.K. was destroyed in 1974 in a huge explosion, killing 28 men. According to chemical engineer Trevor Kletz, perhaps the most important lesson to emerge from the accident was the need to redesign processes so as to reduce the inventory of hazardous substances on site. "If the inventory had been less, the leak [that triggered the explosion] would have been less [he said]. What you don't have can't leak".[1] Kletz also noted that dealing with hazards at the design stage may be cheaper in the long run because "less added-on protective equipment is needed and has to be tested and maintained". Importantly, inherently safe design removes the opportunity for operators to make errors in interpreting or operating add-on safety features.

The term "inherently safe design" is now used well beyond the chemical and process industries and can be defined, generally, as follows: an inherently safer design is one that avoids hazards instead of controlling them.[2]

DOI: 10.1201/9781003629269-5

Two other terms are sometimes used synonymously with inherent safety: passive safety and intrinsic safety. These terms have more specific meanings in particular contexts and so will not be used here.

So how do these ideas about inherently safe design shed light on what went wrong with the 737 MAX?

The design of a new aircraft involves juggling a number of elements, including aerodynamics, propulsion, controls, weight, and structure. These elements make competing demands on the design, and these must be balanced against each other to ensure all components can behave as required. Varying any one of these elements is likely to have impacts on the others, so the balancing must involve a lot of testing and fine-tuning.

Clean-sheet design of an aircraft maximises the chances that potential hazards will be identified at an early stage when they can be avoided, not just controlled. In other words, it facilitates inherently safer design.

The decision to update an existing aircraft design by providing more powerful and fuel-efficient engines has the potential to upset the balance that was originally achieved among the various elements that went into the aircraft design. That is what happened with the 737 MAX.

The original 737 fuselage was located as close to the ground as possible. This allowed baggage handlers to load luggage without requiring much ground support equipment. It also meant that the aircraft could operate more easily at airports lacking aerobridges. Being able to make do with moveable air-stairs helped low-cost carriers keep their costs down – a significant selling point. The 737's engines were mounted under the wings of this low-slung body.

The proposed new engines were larger than the earlier version and would not fit easily under the wings. They would need to be mounted further forward.[3] Clearly, this was a significant change that was likely to upset the balance struck in the initial design. That is indeed what happened. The proposed design of the 737 MAX was "dynamically unstable". According to one knowledgeable commentator, "this is aerodynamic malpractice of the worst kind".[4] According to another, such instability is unacceptable in a transport category aircraft – Boeing was intending to produce the "world's first unstable passenger plane".[5] Using less context-dependent language, the design was not inherently safe.

This was not appreciated by top Boeing people. According to the head of Boeing's commercial airplane division,

> We are going to make this the simplest re-engine possible. We are only going to touch the part of the airplane impacted by the engine and a couple of other improvements.[6]

Such a comment revealed a complete lack of understanding of how the proposed re-engining might impact parts of the aeroplane as far away as the tail.

Sure enough, early testing of a scale model in a wind tunnel revealed a problem caused by the positioning of the proposed new engines – the model had a tendency to pitch up (nose up) during tight, high-speed turns.[7] Commercial airliners would rarely engage in such manoeuvres but there were exceptional circumstances in which they might be necessary. If the problem wasn't fixed, it could lead to a stall. One solution would have been to re-design the tail of the aircraft to *eliminate* the nose-up tendency. This would have been inherently safer. But this would have required other design modifications, and it would have increased costs and delayed the delivery to airline customers. Several decades earlier, in similar circumstances, Boeing had indeed adopted this inherently safer solution,[8] but in the new, commercially-oriented business, that was not considered as a possibility.

The solution adopted therefore was to design some software to *control* the hazard, not *eliminate* it. Attached to the tail of most aircraft are two small wings that together constitute the "tailplane", so-called because it is a flat surface (pair of surfaces) at the tail of the aircraft that *planes* through the air.[9] For many aircraft the tailplane is fixed in place, while for others, the whole tailplane can be tilted upwards or downwards to decrease or increase the lift it generates.[10] The 737 operates in this latter way. This means that the nose of the 737 can be pitched down by tilting the tailplane backwards. The software would do this automatically if the nose-up tendency was detected. Detection would be done by an angle-of-attack (AoA) sensor that measured the angle of the main wings in relation to the oncoming airflow, that is, it measured the steepness of the climb.

The AoA sensor is a tiny vane mounted towards the front of the fuselage, in which position it is very vulnerable to bird strike, or damage from runway equipment while on the ground. The records of the FAA (Federal Aviation Administration) contain hundreds of cases where these sensors have been damaged in this way. The problem was that if this sensor malfunctioned it might result in the software forcing the nose down into an uncontrollable dive. A fundamental principle of aircraft safety is that an aircraft should not be vulnerable to such single point failure. Yet reliance on a single AoA sensor would amount to exactly that. In short, the software solution had introduced a new hazard which in turn would need to be controlled.

Boeing's solution to this problem was as follows. In the normal course of events, aircraft would be fitted with accelerometers that measure G-forces (gravitational forces). The nose-up problem only occurred, potentially, when an aircraft was performing tight, high-speed turns, which is when G-forces are high. So the software could be programmed to activate only if both the accelerometer and the AoA sensor were in agreement. This meant that if AoA sensor alarmed for whatever reason, but the accelerometer did not, the AoA alarm would be assumed to be false, and the software would not activate. This eliminated the single-point failure problem. Boeing went ahead on this basis.

Finally, the first 737 MAX was ready to be test flown. Unfortunately, test pilots soon discovered that the quirk that caused the plane to pitch up in high-speed, tight turns was also present during some tests of low-speed stalls. At this late stage, redesign of the aircraft was out of the question, but the software could be extended to cover low-speed, as well as high-speed situations. This seemed like an obvious solution.

However, engineers were immediately confronted with a new problem. The low-speed scenario in which the nose-up tendency might occur did not involve tight turns, so no excessive G-forces would be involved. The accelerometer alert would never be activated in these situations and so could not function as an independent alarm. It would need to be disabled. The software would now need to fire on the basis of a single input – from the AoA sensor. The 737 MAX design was back to a single-point failure again.

Time was running out and there was no longer any fail-safe mechanism protecting the pilots against erratic software functioning. The pressure on Boeing's engineers to justify reliance on a single sensor was now immense, indeed irresistible, and they duly found various arguments to justify the proposed course of action.[11] One line of argument was that, if the aircraft began to behave erratically as a result of a malfunction of the AoA sensor, pilots would recognise what was happening and hit the cut-out switch on the motor that was controlling the pitch of the aircraft.[12] They would then be able to control the pitch manually.

This argument was problematic, for various reasons. First, Boeing had told the airlines that their pilots would not need any additional simulator training when converting from the 737 to the 737 MAX. Pilots were not even to be informed about the new software operating in the background on the new aircraft they would be flying. Boeing was expecting them to handle an emergency situation where they had little or no idea of what was happening. I address the reasons for this in more detail in a later chapter.

Second, the potential low-speed problem scenarios would be at take-off and landing, when pilots are fully engaged in other activities. To have to deal with an emergency at such a time would be setting pilots up to fail. Unless the pilots responded very quickly, they would indeed be confronted with a full-scale emergency. Once activated, the software was designed to operate for ten seconds, forcing the nose down, after which it would stop and reset. If the pilot brought the nose back up, the software would activate again, and so on, indefinitely. Pilots would find themselves in a battle with an enemy they did not understand. This is exactly what happened in the two crashes. Tragically the test pilots had alerted engineers to this problem of repeated activation, but it had not been dealt with.[13]

There is one last point to be made here. Boeing denied that the engineering solutions they had adopted resulted in a situation where a single point failure could bring down an aircraft. After the Lion Air crash, one of Boeing's vice presidents addressed an audience of pilots on the subject. He said that it was wrong for the press to call it a single-point failure, because "the function [sensor and software] and the trained pilot work side by side and are part of the system".[14] In other words, it took a double failure, that of the sensor and the pilot to bring down the aircraft.

This was a retrospective argument designed to justify a course of action Boeing had adopted. It had not been advanced earlier, when two independent sensors (AoA and G-sensors) had been deemed necessary to avoid the single point failure scenario. It was only now that pilot input was considered sufficient to avoid the single point failure problem.

Interestingly the audience of pilots reacted angrily to Boeing's after-the-fact justification. The doomed pilots had *not* been trained on the operation of the software additions. But even if they had, the logic is unconvincing. Safety in complex hazardous systems depends on engineering designs that are tolerant of errors by front-line staff. A system that is not resilient to human error – where human errors can be fatal – is not a safe system.

So, to sum up, in order to save time and money, Boeing decided to offer a new aircraft to the market, based on re-engining the existing 737, rather than designing a new aircraft from scratch (clean-sheet design). This created a situation where a safety add-on was necessary. Specifically, software was needed that would automatically take control of the aircraft in certain situations for which the 737 MAX was poorly designed. But the software had a fatal flaw – it was highly vulnerable to the failure of one flimsy component – the AoA sensor. In these circumstances, pilots might find themselves fighting a losing battle against the software, with disastrous consequences. The software, designed as a defence against danger, was itself dangerous. Jim Reason, of Swiss cheese fame, wrote about "dangerous defences", nearly 30 years ago.[15] Safety professionals in aviation and other high-hazard industries are well aware of this paradox, but Boeing was in far too much of a rush to attend to such concerns.[16]

NOTES

1 Kletz T, *Learning from Accidents*, 2nd Ed., Butterworth-Heinemann, Oxford, 1994, p. 72.
2 See generally Wikipedia, "Inherent Safety".
3 In addition, the nose gear would need to be raised by about 8 inches, Robison P, *Flying Blind*, Penguin Random House, New York, 2021, pp. 171–172.
4 Travis G, "How the Boeing737 Max Disaster Looks to a Software Developer", *IEEE Spectrum*, first published April 18, 2019, updated, February 3, 2024. https://spectrum.ieee.org/how-the-boeing-737-max-disaster-looks-to-a-software-developer
5 Vidyasagar M, "The Travails of the Boeing 737 Max", http://eceweb.ucsd.edu/~btouri/boeing.html?fbclid=IwAR2nQObiy6tkjuAu9-0l1dNt0DG7XrP6I2EiVTxzbuOJ-oelP5-xNVyqRNs
6 Robison P, *Flying Blind*, Penguin Random House, New York, 2021, p. 207.
7 Robison P, *Flying Blind*, Penguin Random House, New York, 2021, p. 212.
8 Robison P, *Flying Blind*, Penguin Random House, New York, 2021, pp. 30–31.
9 https://en.wikipedia.org/wiki/TailplaneNB.Tailplanedoesnotmeantailoftheaeroplane.The trailingedgeofthetailplanehasmoveablesurfacescalledelevators.Theseprovideanadditional way to control the aircraft pitch. See https://aviation.stackexchange.com/questions/21114/why-do-some-fighter-jets-have-movable-horizontal-stabilizer-instead-of-elevators
10 A moveable tailplane is called a stabilator.
11 See Gates, "The Inside Story of MCAS", *Seattle Times*, pp. 12–13/30.
12 Robison P, *Flying Blind*, Penguin Random House, New York, 2021, p. 244.
13 Robison P, *Flying Blind*, Penguin Random House, New York, 2021, p. 245.
14 Robison P, *Flying Blind*, Penguin Random House, New York, 2021, p. 294.
15 Reason J., Chapter 3 of *Managing the Risks of Organizational Accidents*.
16 See Gates D, "The Inside Story", *Seattle Times*, June 22, 2019.

6 Human Factors

So far, we have focussed on the way Boeing's design decisions contributed to the emergency situation in which pilots found themselves. We need now to consider the actions of the pilots as the emergencies developed.

Boeing's position was that "the flight crew has the final authority over the operation of the airplane" and can over-ride the plane's automatic actions.[1] Boeing claimed that they had provided pilots with all the information they needed to control a nose-down crisis, such as happened in both crashes, and had reinforced this message to airlines after the first crash. "We provide all the information that is necessary to safely fly our planes", said the CEO.[2] The U.S. National Transportation Safety Board (NTSB) agreed, at least in the case of the Ethiopian Airlines crash, that Boeing had indeed provided this information.[3]

Boeing went further. The clear implication in what it said was that American pilots would have known what to do; the problem lay with the Indonesian and Ethiopian pilots and the airlines for which they flew.[4]

Where it is found that the behaviour of the pilots contributed to an accident, this should not be seen as the fault of the pilots. If that is where an accident analysis culminates, the goal of accident prevention is no further advanced.

The question that must be addressed is *why* the pilots behaved as they did. Given that the pilots had apparently been provided with the necessary information about what to do, why did they fail to follow instructions? If we can answer these questions, then lessons will emerge which, if implemented, will improve the safety of aviation. That is what the discipline of *human factors* is all about. This is well understood by many accident investigation agencies around the world.[5] Importantly, the discipline is concerned not only with why people make mistakes but also with why they are sometimes able to save the day, as did Captain "Sully" when he landed his aircraft on the Hudson River (see later).

In what follows I shall focus on the second MAX crash – the Ethiopian Airlines crash – primarily because the controversy that surrounded the main accident report, produced by the Ethiopian Accident Investigation Board, generated additional information about the subjective experience of the pilots. This report identified the software problems discussed earlier, but it did not explicitly list pilot error as a cause or one of the factors contributing to the accident.[6] No explanation is provided for this omission, and we can only surmise that the reason may have been to protect the Ethiopian Airlines or governmental authorities from any criticism. However, both the U.S. NTSB (National Transportation Safety Board) and the French BEA (Bureau of Enquiry and Analysis)[7] were involved in the accident investigation. When the official Ethiopian report was issued, the French and American investigators were clearly furious that, despite their input, the final report appeared to downplay the role of human factors. Accordingly, they both took the almost unprecedented step of issuing statements critical of the official report and filling in some of the missing information. The following account relies heavily on the French statement.

DOI: 10.1201/9781003629269-6

As is standard practice at take-off, the Ethiopian plane was being flown manually, that is, without the autopilot operating. Almost immediately after take-off, the aircraft apparently hit a large bird which damaged the angle-of-attack (AoA) sensor. The damaged sensor began to give false readings, indicating that the nose was too high, and the aircraft was about to stall. Thereupon, the control column (joystick) began to shake, as it was designed to do, to alert the pilot to the presumed impending stall.

An experienced airline pilot, David Walker, who has studied these events closely makes the following comment[8]:

> Every airline captain in the world would be alarmed at having the stick shaker activate four seconds after take-off. As an experienced airline pilot, my blood ran cold when I read this in the report. I've *never* had a stick shaker sound continuously outside a simulator. This event would have received the captain's full attention…

The stick shaker continued for the remainder of the flight, falsely, but insistently, warning of an imminent stall. There is a standard procedure to be applied in the event of a stall warning, which crews are supposed to know, without needing to refer to a manual. It requires the pilot to hold the stick firmly, disengage the autothrottle and the autopilot (if it is engaged, which it was not at this point), and smoothly apply a nose-down input to increase speed and avoid the stall. Only the nose-down input was performed by the flight crew. The autothrottle was not disengaged. It remained at full take-off power for the rest of the flight. This resulted in the aircraft reaching a dangerous speed, which made it increasingly difficult for the crew to control the plane during the remainder of its short flight. There is evidence that the flight crew was not aware either of the speed that the aircraft reached or that they needed to disconnect the autothrottle to reduce speed. The French statement emphasises that the failure to disengage the autothrottle, and the lack of crew discussion about this, "played a role in the chain of events that led to the accident".[9]

Again, Walker expresses a somewhat contrary view about this matter:

> Airline pilots learn to increase thrust when they are in doubt. Under no circumstances could I imagine myself reaching for the throttles and reducing thrust while the stick shaker was doing its thing, particularly not right after take-off while operating close to the ground.

On this view, the pilot was quite entitled to ignore the standard instruction at least for the time being. The problem is that he did not throttle back as the speed increased, perhaps because of the confusion being experienced in the cockpit.

At a height of approximately 350 feet, the captain tried, unsuccessfully, to engage the autopilot, although the airline policy requests pilots to wait until they reach 500 feet to do this. According to the French statement, "This premature action…. may be symptomatic of a state of stress that had been rapidly developing following the activation of the stick shaker … immediately after take-off." In any case, given the earlier mentioned requirement to disconnect the autopilot if the stick is shaking, the captain should not have been trying to engage the autopilot at this stage.

The Ethiopian report raises another possibility – that the pilot had in mind a contradictory instruction in the flight crew training manual that "when a non-normal situation occurs, the maximum use of auto flight system is recommended to reduce workload."[10] Walker concurs:

> Already, only 46 seconds from the start of take-off, there is a lot of chaos and confusion for the captain to manage. He does the smart thing and decides to shed some of his workload: He attempts to engage the autopilot. This is a wise decision; pilots are taught to reduce their own workload when encountering any sort of problem, so they are better enabled to manage bad situations. Manually flying an aircraft is cognitively demanding.

In short, the standard procedure for dealing with a stall warning was problematic when that warning occurred immediately after take-off. Indeed, it was contrary to other instructions and training. This inconsistency could only have exacerbated the situation.

So far, I have avoided giving a name to the rogue software, to minimise the use of acronyms. But this will become increasingly clumsy, so I now introduce its name: – Manoeuvring Characteristics Augmentation System, or MCAS, for short, pronounced Em-CAS.

The important point to note here is that the problem the pilots were having at this stage had nothing to do with MCAS, which had not yet been activated. MCAS could be triggered by the AoA sensor if two conditions were satisfied: (i) the aircraft was under manual control, which it still was, and (ii) the wing flaps, extended for take-off, had been fully retracted for normal flight; this had not yet happened. So the roller coaster ride had not yet begun.

The pilot made a second attempt to engage the autopilot six seconds after the first, with the aircraft now above 500 feet. This, too, failed, prompting the captain to say: "what's going on?" The French report comments that "the captain's question remained unanswered and did not trigger any process of information acquisition, cross-check or crew decision making". None of the accident investigation reports appear to have answered the question.

The French report notes that, to this point, and indeed throughout the whole flight, "the coordination and the communication between the captain and the first officer was [were] very limited and insufficient. There was neither discussion nor diagnosis with respect to the nature of the events on board. The situational awareness, problem-solving, and decision-making were therefore deeply impacted. The first officer's lack of proactivity, which comes out from the cockpit voice recorder transcripts, seems to show that he was overwhelmed by the events on board from the moment the stick shaker triggered. His low flight experience (300 hours total) may have accounted for this situation."

At 1,000 feet a third attempt to engage the autopilot was successful. But 33 seconds later, for some unexplained reason, the autopilot disconnected, automatically. The aircraft was now back to manual flight which meant that, as soon as the flaps were fully retracted, MCAS sprang into action, took control, and issued a strong nose-down command. The captain struggled to reverse this command by pulling hard on the stick. The stick controlled the elevators – moveable surfaces on the trailing edge of the tailplane. Using almost as much force as he could muster, he was able to return the plane almost to level flight.

According to the French report, "During this phase, the physical efforts applied by the crew on the column probably impacted their situational awareness and their cognitive resources and did not allow them to undertake *the proper actions*." (emphasis added)

Boeing implicitly acknowledged this problem. In a subsequent software update,[11] the company promised that in future the software will never be able to exert greater force "than can be counteracted by flight crew pulling back on the column. The pilots will continue always to have the ability to override the software and manually control the airplane":

The "proper actions" that the pilots should have followed included making use of a secondary system for adjusting the nose-up/nose-down position of an aircraft – trimming.[12] In the case of a 737, trimming involves making adjustments to the pitch (AoA) of the tailplane. For the 737 this can be done electrically by pushing a button on the control column.[13] This would have reduced the need to pull back on the control column with such force. But the crew failed to make adequate use of this option.

The French investigation team ran simulator sessions to see what its pilots would have done in a similar situation. Its report comments that the simulator crews "felt it was instinctual to use as much electric trim as needed" to cope with the input from the errant software. But they also recognised this "was not very common in a normal flight". This is a disturbing comment. Does it mean that in the final analysis safety depends on pilot instinct? And if this is the conclusion reached in the safe environment of a simulator, can pilots in a real emergency be expected to be able to rely on such instincts?

The crew did manage to pull the nose up a second time, only to have the MCAS activate a third time, pushing the nose down. Again, the crew pulled together with all their might on the control column. The high speed of the aircraft was now creating such extreme air flow forces on the adjustable surfaces on the tailplane that the pilots could no longer control the aircraft. Sounds of exhaustion and shortness of breath were heard on the cockpit voice recorder.[14] A fourth MCAS activation sent the aircraft nose-first into the ground at well over 900 kph.

This highlights another problem: the MCAS fired four times before it became entirely unmanageable. The subsequent software update acknowledged this problem and promised that henceforth the software "will only provide one input for each elevated AoA event. There are no known or envisioned failure conditions where MCAS will provide multiple inputs", as happened in the two crashes. If a crew manages to deal with one uncommanded nose-down event, it will no longer be at risk of subsequent escalating events.

If we take a step back from the specific details, the subjective experience of the pilots comes more clearly into view. As one commentator put it:

> The crew was increasingly distracted by warning lights and messages on their instrument panel; the stick shaker, indicating the jet was too slow; a loud clacking noise indicating it was too fast; a robotic voice declaring "Don't sink! Don't sink!" which was a warning the jet was close to the ground.[15]

These alarms would have generated a state of sensory overload, in which the brain is no longer capable of processing all the input coming from the senses. The whole system of warnings and alerts was undermined by the failure of designers to think carefully about how pilots might react to these cascading alerts in an emergency.

To summarise this point, the cockpit of the 737 MAX was *not* an error-tolerant environment. It did *not* forgive. Each error led to the next, escalating steadily towards a disastrous outcome. The software modifications made after the accidents provide a far more error-tolerant system. In its rush to get its plane to market, Boeing had not paid adequate attention to human factors in its design of the MCAS system. Only after these accidents did it do so.

There is another statement in the software alert that should be highlighted, indeed shouted from the rooftops. The whole sequence of events was initiated by a false reading from the single AoA sensor. There were in fact two AoA sensors, one on either side of the fuselage. In the standard version of the aircraft, acquired by Ethiopian Airlines, only one was connected to the MCAS and likewise to the stall alert. There was a safer system, which Boeing offered as an optional extra, at considerable additional cost. This system used both AoA sensors and compared inputs from the two. If there was significant disagreement between the two readings, MCAS would not activate, and nor would the stall alert. In the subsequent software update, Boeing implicitly recognised that the use of only one sensor made the plane vulnerable to a single-point failure: it announced that in future all 737 MAX planes would use both sensors. This was dramatic recognition by Boeing of its contribution to the two accidents.

The Expert Panel commissioned by the FAA after the accidents highlights the role of human factors in the two accidents, as follows:

> [In the 1970s and 1980s] Boeing's human factors specialists played an integral role in enhancing the design and functionality of the airplanes' flight decks. Boeing's human factors in flight deck design and operations were the gold standard.... Since then, the role of human factors and its influence eroded due to a series of administrative decisions at Boeing, which includes reorganization, decentralization, downsizing, and relocating the company's headquarters.[16]

In short, Boeing's lack of attention to human factors is yet another consequence of its transformation into a profit-driven company.

FAILURE TO FOLLOW PROCEDURES

We have yet to inquire into Boeing's claim that it had adequately informed pilots of what to do in a nose-down emergency, or as Boeing describes it, a non-normal situation. It is true that Boeing had issued a "bulletin" to all airlines, including Ethiopian Airlines, after the Lion Air crash. The aim of the bulletin was to emphasise that the situation encountered by the Lion Air crew was already covered in the procedures Boeing had provided to deal with non-normal situations. All that crews need to do is follow those procedures. No specific mention was made of MCAS, although it was acknowledged that "repetitive cycles of uncommanded nose down" might occur if crews did not follow the procedures. Airlines were instructed to insert the bulletin into their Flight Crew Operations Manuals. One former NTSB investigator observed that the bulletin should have been "much more clear and transparent and had a higher sense of urgency". According to another critic, the "the procedures [i.e. the bulletin] were clearly written to minimise any sort of liability" on Boeing's part.[17]

Interestingly, Ethiopian Airlines had concerns about the bulletin before the accident and emailed three questions to Boeing. A Boeing spokesman declined to answer the first two on the grounds that they were about the Lion Air crash which he could not comment on. The last question was answered by again drawing attention to existing procedures. Moreover, the layout of the letter was sloppy, and it was difficult to identify the text of the answer. It was truly a failure as a piece of communication.[18] Relatives of some of the victims highlighted these issues in criminal proceedings against Boeing in 2024, arguing that if Boeing had provided more fulsome answers, the pilots might well have been able to avoid the accident.[19]

In addition to the Boeing bulletin, Ethiopian Airlines, along with others, received an "emergency airworthiness directive" from the FAA, which paralleled the advice given by Boeing. The directive also used language that tended to downplay the issue, describing it as an "unsafe condition" that could cause "possible impact with the terrain".

Ethiopian Airlines made these documents available to individual air crew on their iPads (logipads). Clearly, they needed to be carefully studied, but the airline made no provision for ensuring that crew read, understood, or absorbed this information. Nor was any training provided. According to the French report, this failure to ensure that crews understood the information provided was a factor contributing to the accident.[20]

One limitation of Boeing's procedures is particularly relevant here.[21] They are written to be applicable to any nose-down emergency, independently of how it originated. In the case of the Ethiopian crew, they were already dealing with one crisis – a stall warning occurring very close to the ground – when they were confronted with a second – an uncommanded nose-dive towards the ground. The use of full throttle to deal with the first crisis made the second, unmanageable. The scenario of distinct, but linked, crises that the pilots had to deal with was not explicitly covered by Boeing's procedures. Perhaps it is unrealistic to expect procedures to do so. All the more reason not to have relied on a single flimsy sensor, the failure of which triggered both crises.

Be this as it may, it is difficult for pilots in the midst of a fast-moving crisis to recall training they may have had, unless that training has been extremely intensive, and the required actions are semi-automatic. Moreover, if pilots are unsure of what to do in a highly stressful crisis situation, it is unreasonable to expect that they will be able to hunt through a procedure manual, find the right procedures, and apply them methodically.

The gold-standard training nowadays is on simulators – mock-up aircraft cockpits which simulate the reality of piloting an aircraft in a remarkably faithful way. In particular, simulators can simulate a variety of crisis situations which pilots need to be trained for. There are different simulators for different aircraft types. Pilots undergo refresher training in simulators every 6–12 months. These refreshers may also simulate the circumstances of recent aircraft disasters. If Ethiopian pilots had been put through a simulated Lion Air scenario, this might well have saved the lives of 157 people.

Simulator training is expensive and Boeing had promised its airline customers that pilots would not need any simulator training on the 737 MAX beyond what they had already received on the previous version of the 737. This promise was an important selling point for Boeing in its quest for a commercial advantage over its main competitor, Airbus. The two crashes demonstrated just how short-sighted Boeing's promise had been.

The Ethiopian accident investigators treated Boeing's failure to make available simulator training for the 737 MAX as one of the contributing factors to the accident.[22] Interestingly, Boeing implicitly accepted this criticism. In the previously-mentioned software update, the company committed to a comprehensive training program, *including simulator training*, for new pilots seeking certification to fly the MAX. For pilots upgrading from the previous version of the 737, this would include specific training on MCAS.

Boeing's implicit suggestion that the Ethiopian accident (and the Lion Air crash before it) were ultimately attributable to the pilots' failure to follow instructions was monumentally unfair. The captain of the doomed Ethiopian aircraft was well qualified and well respected. His training record included glowing praise from arm's length training captains and check pilots.[23] The situation he found himself in was impossibly stressful. If Boeing had carried out a thorough human factors analysis during the development of the 737 MAX, the pilot would never have been placed in that situation. The failure adequately to consider human factors must be considered one of the fundamental causes of both accidents.

THE DANGERS OF AUTOMATION

For the 737 MAX, automation had solved one problem – the nose-up tendency of the MAX, but had spawned another – the possibility of repeated nose-down commands that the pilots would be unable to counter. Boeing had made an assumption that if the automated systems failed for any reason, pilots would be able to take control of the situation and counteract the failure. In fact, the pilots were unable to do this. Boeing's assumption was wrong.

The ultimate promise of automation is that it will be able to cope with every conceivable emergency and that planes will be able to fly without pilots. Boeing is moving in this direction with cargo planes, and it hopes, someday, to make passenger planes that will fly, with little or no role for pilots.[24]

Here is what its CEO said in November 2019, "We are going to have to ultimately almost – almost – make these planes fly on their own."

Fully autonomous, driverless cars are now a reality. A driverless car can coast to standstill, if need be, in an emergency; that is not an option for a pilotless plane. The airline industry will have to proceed very much more cautiously, given that the stakes are so much higher.

For the foreseeable future, pilots will need to be able to take control in order to prevent malfunctioning automated systems from driving an aircraft to destruction. As Meshkati has noted, "when something goes wrong … you need the human operator to come back into the loop to solve the problem.[25] A former Boeing and Airbus pilot expressed this idea in uncompromising language, "no flight-control system that is driven by a computer should go without either a manual override or way to engage a complete computer disconnect in the event of complete runaway".[26]

Boeing has learnt from the 737 MAX crashes and re-programmed the MCAS software so that "it will never be able to provide more input that the pilot can counteract using the control column alone". For this reason, if for no other, we are unlikely to see an exact repetition of the accidents that brought down the Lion Air and Ethiopian aircraft.

But there is a more general problem lurking along the way to 100% automation. As time goes by there will be less for pilots to do in the cockpit and less opportunity to use their skills. This may lead to an erosion of pilot skills, leaving them unable to deal with serious incidents when they arise. Moreover, according to one industry expert, the challenge for pilots is "to maintain situational awareness, not only as to where you are and how your airplane is flying, but what the heck the automation is doing, and when you need to shut it off and hand fly the aircraft."[27] Automatic flight can be in one of several modes – take-off, steady state, etc. Pilots sometimes lose awareness of what mode the aircraft is flying in – so-called "mode confusion". This has been implicated in various automation-related accidents. So here is the paradox. The rarer it is that pilots need to over-ride computer systems, or take over in an emergency, the greater the likelihood that they will be unable to cope with crisis situations when they occur.

The most effective way to ensure that pilots have the skills to deal with these rare but potentially catastrophic events is to train specifically for them. Air force pilots in peacetime spend much of their time training for the extreme situations that may occur in real operations. Commercial airline pilots, however, do refresher training in flight simulators on the ground relatively infrequently – every 6–12 months. This is not enough to ensure pilots know how to react in a crisis situation, and they often find it necessary to resort to flight manuals.

Consider this example. A Southwest airline 737 MAX was taking off when its left engine ingested a large bird, suffered severe damage, and caught fire.[28] The left engine provides fresh air to the cockpit, so smoke began pouring into the cockpit. The pilots had to don their oxygen masks and smoke goggles. Pilots regard smoke in the cockpit as "one of the top five bad emergencies you can have in an aircraft." In this case the pilots declared an emergency. The control tower asked if emergency vehicles should be deployed and the response was "we need everything you have" – a response that gives some inkling of the stress they were under. Air traffic control invited them to land immediately, but the captain replied that he was "running some checklists" and needed a few more minutes. When asked a second time, again he replied that "We aren't quite ready." The problem was that there were three different checklists of actions potentially relevant to the situation. Eventually, the pilot worked out what they needed to do, and the plane landed safely. The pilots had got it right. But in an emergency in which the pilots need to switch off the automation and take control of the aircraft, there may not be time to read a manual. That is when everything will depend on the training and expertise of the pilot.

A good example of this is the famous emergency landing on the Hudson River in 2009.[29] The plane had just taken off from La Guardia airport in New York when it hit a flock of geese. Both engines were destroyed, but captain "Sully" Sullenberger was able to glide the plane to a safe water landing. He had no time to consult manuals. He had been taught in the classroom how to carry out a water landing, but had never trained for it – there was no simulator training for this. On the other hand, he had trained, and then served, as an air force fighter pilot as a young man, and subsequently spent 30 years flying commercial airliners. All this training and experience took over. As he said,

> One way of looking at this might be that for 42 years, I've been making small, regular deposits in this bank of experience, education and training. And on January 15, 2009, the balance was sufficient so that I could make a very large withdrawal.

Was this a failure of automation? Certainly it shows the limits of automation at the present time. Importantly, it suggests the level of skill that might be necessary to deal successfully with an automation-related emergency.

Automation itself is not necessarily a danger; in fact airline travel is now safer as a result of progressive automation. The real danger is that companies may see automation as an opportunity to cut pilot training. Experts have been warning about this for years. According to one, the financial savings to be made from reduced crew training means that, even after automation-related crashes, the industry "will return to the comfortable assurances by all concerned that any real issue can be easily solved by simply increasing the levels of available automation ...... Lured by the false promises that automation turns lead into gold, airline managers make the manuals ever thinner and the training ever shorter, ... [Airlines rely] more upon computer-based training to minimum standards, and less upon a determination of whether flight crews have actually learned enough to safely and efficiently operate the airplanes to which they are assigned".[30]

So in the end, the issue of automation is part of the broader topic of human factors. The introduction of automation does not eliminate the need to consider human factors; it heightens it.

NOTES

1 Baker S, "The Boeing 737 Max Crashes Have Revived Decades-Old Fears about What Could Happen When Airplane Computers Become More Powerful Than Pilots", *Business Insider*, February 17, 2020,. https://www.businessinsider.com/boeing-737-max-fatal-crashes-revive-fears-automation-planes-2020-2
2 Robison P, *Flying Blind*, Penguin Random House, New York, 2021, p. 282.
3 NTSB comments on draft Ethiopian Accident Investigation Board report on Ethiopian Airlines crash, March 10, 2019, Section 1.4. https://www.ntsb.gov/news/press-releases/Pages/NR20221227.aspx
4 Robison P, *Flying Blind*, Penguin Random House, New York, 2021, p. 379.
5 See the NTSB comments on Draft Ethiopian report, Section 2. "As we have reiterated throughout the investigation, design mitigation must adequately account for expected human behaviour to be successful, and a thorough understanding of the flight crew's performance in this accident is required not only for robust design mitigations but also for operational and training safety improvements necessary to achieve multiple layers of safety barriers to trap human errors and prevent accidents". p. 4. https://www.ntsb.gov/investigations/Documents/US%20comments%20ET302%20Report%20March%202022.pdf

6 Ethiopian Accident Investigation Board Report on Ethiopian Airlines Crash, p. 255. https://context-cdn.washingtonpost.com/notes/prod/default/documents/18b9f943-d5ca-4143-898f-2596bdd1286c/note/8ba51389-8298-4c08-8ef6-6416b262fd1b.pdf
7 Bureau d'Enquêtes et d'Analyses pour la Sécurité de l'Aviation Civile, the French Civil Aviation Accident Investigation Bureau.
8 Walker D, "Accident Timeline of Ethiopian 302", *Medium*, June 23, 2019, https://medium.com/@dawalk6/accident-timeline-of-ethiopian-302-ed20e33f900a
9 See paragraph on "probable cause" and "contributing factors" in BEA report.
10 Ethiopian Accident Investigation Board Report on Ethiopian Airlines Crash, p. 221.
11 https://www.boeing.com/commercial/737max/737-max-update/737-max-software-updates#overview
12 https://skybrary.aero/articles/trim-systems#:~:text=Description,attitude%20without%20any%20control%20input
13 If electric system fails, or is switched off, the adjustments can be made by manually rotating a large "trim wheel" in the cockpit.
14 Ethiopian Accident Investigation Board Report on Ethiopian Airlines Crash, p. 250.
15 https://www.seattletimes.com/business/boeing-aerospace/final-report-on-boeing-737-max-crash-disputed-agencies-note-pilot-error-as-a-factor/#
16 ExpertPanelSafetCultureReportopcit.,pp.25–26.https://www.faa.gov/regulationspolicies/rulemaking/committees/documents/section-103-organization-designation
17 "Final Report on Boeing 737MAX Crash Sparks Dispute Over Pilot Error", https://www.seattletimes.com/business/boeing-aerospace/final-report-on-boeing-737-max-crash-disputed-agencies-note-pilot-error-as-a-factor/#
18 Ethiopian report, p. 289.
19 https://www.nytimes.com/2024/10/09/us/politics/boeing-ethiopian-airlines-crash-questions.html. See also https://www.nytimes.com/2024/12/05/us/politics/boeing-max-guilty-plea-rejected.html
20 For relevant procedure see Ethiopian Accident Investigation Board Report on Ethiopian Airlines Crash, pp. 168–169. For required responses see pp. 164–167.
21 The Ethiopian report also draws attention to this at p. 225, contributing factors 3,4,5.
22 Ethiopian Accident Investigation Board Report on Ethiopian Airlines Crash, p. 255.
23 Walker D, "Accident Timeline of Ethiopian 302", *Medium*, June 23, 2019.
24 https://www.wsj.com/articles/max-crashes-strengthen-resolve-of-boeing-to-automate-flight-11577816304
25 Baker S, "The Boeing 737 Max Crashes Have Revived Decades-Old Fears about What Could Happen When Airplane Computers Become More Powerful Than Pilots", *Business Insider*, February 17, 2020, https://www.businessinsider.com/boeing-737-max-fatal-crashes-revive-fears-automation-planes-2020-2
26 "Deadly Assumptions… https://www.nytimes.com/2019/06/01/business/boeing-737-max-crash.html, p. 6.
27 Baker S, "The Boeing 737 Max Crashes Have Revived Decades-Old Fears about What Could Happen When Airplane Computers Become More Powerful Than Pilots", *Business Insider*, February 17, 2020/. https://www.businessinsider.com/boeing-737-max-fatal-crashes-revive-fears-automation-planes-2020-2
28 Gates D. "After 2 serious 737 MAX engine incidents at Southwest, Boeing alerts pilots", *Seattle Times* June 18, 2024 https://www.seattletimes.com/business/boeing-aerospace/after-2-serious-737-max-engine-incidents-at-southwest-boeing-alerts-pilots/?utm_sourc%E2%80%A6
29 "Sully Sullenberger", Wikipedia. https://en.wikipedia.org/wiki/Sully_Sullenberger
30 Baker S, "The Boeing 737 Max Crashes Have Revived Decades-Old Fears about What Could Happen When Airplane Computers Become More Powerful Than Pilots", *Business Insider*, February 17, 2020,. https://www.businessinsider.com/boeing-737-max-fatal-crashes-revive-fears-automation-planes-2020-2

7 Certification of the MAX

The 737 MAX could not have gone into service without the approval of the regulator, the FAA (Federal Aviation Administration), which was duly given in 2017. How could the FAA have certified as safe an aircraft that used a patently unsafe system, such as MCAS? That is the question to be explored in this chapter. As we shall see, this was a classic case of "regulatory capture", the capture of the regulator by an organisation it is supposed to regulate.

Most discussions of how the FAA came to certify the MAX, despite its unsafe MCAS, implicitly blame the regulator of lacking the moral fibre to stand up to Boeing, and of "cozying up" to Boeing. These judgments tend to personify the FAA in a way that hinders understanding. In fact, many of the FAA's failings must be seen in a broader context and are matters for which Congress must accept responsibility.

But it is also true that Boeing seriously misled the regulator. Among other things, this chapter will explore how and why the regulator was misled. It will highlight aspects of Boeing's organisational structure and functioning that allowed knowledge of the true state of affairs with MCAS to remain fragmented within Boeing. This meant that no Boeing employee had the full picture, and no one was fully accountable for the company's dealings with the FAA. The idea that Boeing intentionally deceived the regulator will be dealt with in more detail in Chapter 8.

The certification process occurred in stages, over a five-year period. At the outset, Boeing asked the FAA to accept that the basic MAX design was the same as all the 737 variants that had gone before, dating back to the first 737, granted certification in 1967. In the FAA language, Boeing was asserting that the MAX was the same aircraft *type*. Assuming the FAA accepted that proposition, the MAX would not need to undergo the full new design approval process. It would need only a "supplemental-type certificate" which involved a less onerous review.

Boeing had another, vitally important reason to convince the FAA that the MAX was just another 737-type aircraft. It meant that pilots who had flown the previous 737 version would not need expensive simulator training. This was a major selling point for airlines considering placing orders for the MAX. Boeing had backed this with a promise to Southwest Airlines in 2011 that if the FAA ultimately required simulator training for pilots, Boeing would give the airline a $1-million-per-plane rebate. Southwest had placed 200 firm orders for the MAX with options on nearly 200 more, so if simulator training were to be mandated, that would cost Boeing somewhere between $200 million and $400 million.[1] This provided Boeing with a powerful incentive to do everything it could to convince the FAA that the MCAS system was not a significant departure from the original 737 design.

In 2013, Boeing employees formulated a plan to help avoid increased "cost" and "greater certification and training impact", by describing MCAS as an addition to an existing stability system, rather than a new function.[2] This was a blatant attempt to divert FAA scrutiny. If dishonesty is the issue, this is where it began.

DOI: 10.1201/9781003629269-7

With this background we can begin to address the ways in which regulatory oversight failed. The FAA had insufficient staff to ensure Boeing was complying with FAA rules and acting in good faith. Accordingly, it had progressively devolved this role to Boeing employees themselves. It had designated 1,500[3] of Boeing's employees as "authorised representatives" to act on its behalf, particularly with regard to certification of the many different systems that would go into the MAX. In the early days of this delegation procedure, authorised representatives were selected by the FAA, but the procedure evolved, so that by 2009 before the Boeing had embarked on the MAX program, these FAA representatives were selected by Boeing, and were only vaguely under FAA supervision.[4] These people had regular functions within Boeing, as well as their FAA supervisory function. This meant that at times they were hopelessly conflicted. One of the investigations after the two crashes identified a company engineer who worked on a particular design and then switched hats to approve the design on behalf of the FAA.[5] A more compromised situation is hard to imagine.

Boeing's surveys showed that almost 40% of the authorised representatives had been in situations in which they perceived potential "undue pressure" in their performance of their FAA role.[6] Here is a particularly clear example of this problem. One of the FAA authorised representatives was present at the above-mentioned 2013 meeting about how to describe MCAS to the FAA.[7] That person concurred with the decision to downplay the significance of MCAS. In other words, the FAA's representative failed to represent the interests of the FAA.

The problem might not have been so acute if the FAA had exercised effective oversight of its authorised representatives. But it didn't. There were indeed personnel at FAA headquarters designated to exercise some degree of oversight over the certification work of FAA's authorised representatives on site. But they were engaged in certification work themselves. In reality, they prioritised their own certification work and allowed their oversight activities of FAA representatives on site to fall far behind schedule. At one point, a third of their planned oversight activities were more than 430 days late, that is, well over a year behind schedule.[8]

Over time, FAA engineers delegated an increasing amount of the 737 MAX certification work to Boeing employees acting as authorised representatives – as much as 87% near the end of the MAX certification process.[9] At this point, then, the certification of the 737 MAX was, to a large extent, self-certification by Boeing.

This evolution was partly at the behest of the aerospace companies, who had for years complained to Congress about the delays in certification. Congress responded in 2003 by ordering the FAA to delegate more nuts-and-bolts compliance work to the plane manufacturers themselves.[10] This is one of the reasons that Congress must be held ultimately responsible.

The FAA has pushed back against criticisms that it had been captured by the industry, justifying the outsourcing of certification as follows: "the ever-expanding magnitude of the U.S. aerospace industry requires that the agency delegate an increasing number of oversight functions".[11] That much is true. But the agency was not willing to recognise what this meant. After the second MAX crash, it said, "the FAA has never allowed companies to police themselves or self-certify their aircraft".

It further explained, "with strict FAA oversight, delegation extends the rigor of the FAA certification process to other recognised professionals [Boeing employees], thereby multiplying the technical expertise focused on assuring an aircraft meets FAA standards".[12] This was a fine statement of the theory, but a complete denial of the reality.

The FAA's delegation of the certification process to the industry is a direct result of inadequate resourcing of the regulator. It is worth pointing out that this is exactly what happened to the Canadian air safety regulator in decades past. In Chapter 10, I shall say more about a report by the Canadian Judge Moshansky who headed the inquiry into an aircraft crash at Dryden, Ontario, in 1989. Here is what he said in 2017.

> The lack of adequate funding of Transport Canada's Regulatory Oversight Branch in the 1980s was the root cause of the Dryden crash in 1989, and it remains hanging like a Damocles Sword over the Canadian air travelling public today[13]

This remark was made to a committee of the Canadian parliament and he concluded that this should be treated as a matter of urgency. "Among the lives you save could be your own", he said. "Following Moshansky's reasoning, we might conclude that the inadequate funding of the FAA was the root cause of the 737 MAX crashes. The idea of root cause is an imprecise metaphor, and I shall say much more about causation in Chapter 11. But Moshansky's point is that air safety depends ultimately on the effectiveness of the regulator.

To sum up this point, the House committee finding on this topic was that "excessive FAA delegation to Boeing has eroded FAA's oversight capabilities".[14] Stronger language would have been justified. The outsourcing of government regulatory responsibilities to private sector agencies has been implicated in many disasters and scandals. The FAA system of outsourcing to the staff of the regulated entity itself is a particularly egregious example.

A CLOSE-UP OF THE MCAS CERTIFICATION PROCESS

We focus more sharply, now, on the process which culminated in the certification of MCAS. Timing of events becomes important here.

In **January 2016**, Boeing began testing a prototype of the new aircraft. As described in Chapter 3, the test pilots quickly discovered that the aircraft performed poorly at low speeds, risking a stall.

To recap, engineers chose to solve this problem by extending the capabilities of the MCAS software to cover the risk of low-speed stall. However, the low-speed scenario did not involve tight turns, so no excessive G-forces would be involved. The accelerometer would never be activated in these situations and so could not function as an independent stall alarm. It would need to be disabled. The software would now fire on the basis of a single input – from a single angle-of-attack (AoA) sensor.

In **March 2016**, two senior Boeing managers, one of whom was the chief project engineer for the MAX, approved this extension of the MCAS system.[15] The latter said subsequently[16] he didn't know at the time that

1. MCAS would activate on a single AoA alert;
2. Boeing had internal test data showing one test pilot took more than ten seconds to respond to an erroneous MCAS activation, in a simulator.[17]

The significance of this last point is two-fold. First, FAA guidance was that pilots should be able to respond in four seconds. Second, a delay of more than ten seconds made a crash inevitable.

Knowledge of the test pilot's failure to respond for more than ten seconds was quite widespread within Boeing. Indeed four FAA-authorised representatives were aware of this issue but failed to report it to the FAA.[18] This is another dramatic example of the failure of the FAA's system of delegated responsibility.

The chief project engineer defended his lack of awareness of these matters by saying that he relied on the advice of engineers in the MCAS project. This is a critical communication failure that immediately raises the question of his accountability and that of the engineers who advised him. How could this have happened?

The chief project engineer explained it in this way. Although he was responsible for signing off on key decisions on the MAX, he did not actually supervise any engineers. "You could say that none of them worked for me, but all of them worked for me", he said. The House report commented that "this reporting structure contributed to the overall lack of accountability on the MAX program".[19] To put it bluntly, this was an organisational arrangement that meant that no one was truly accountable.

The House report further notes that Boeing thanked the chief project engineer for keeping to the MAX's production schedule, by rewarding him with restricted stock options after the airplane's first flight in 2016.

Also in **March 2016,** Boeing's *chief technical pilot* sought and obtained FAA approval to remove all reference to MCAS from the flight manual.[20] His argument was that pilots had no need to know about the system because it only operated in extreme circumstances, far outside normal operating conditions.[21] This was quite wrong. The *chief project engineer* had just approved the extension of MCAS to low-speed situations, which were far less extreme. The *chief technical pilot* later said that he had no knowledge at the time of the expanded capability of MCAS that had just been approved.[22] The FAA's approval was a big win for Boeing because it supported its claim that no additional simulator training would be necessary.

In **August 2016**, the FAA finally and formally accepted that MCAS would not require simulator training.[23] This was the crowning achievement of Boeing's whole certification campaign.

Then, in **November 2016,** came one of the most significant events in this chronology. Boeing handed over a safety system assessment of MCAS to the FAA for final checks, as it was required to do. However, the documents handed over related to an earlier version of MCAS – before its function had been expanded to cover the possibility of low-speed stalls! The safety assessment said nothing about the changes made in subsequent MCAS versions which had made it far more dangerous.[24] The FAA certification engineers accepted the submitted MCAS safety assessment, without knowing anything about the final, expanded version. This was truly a low point in the sequence of events leading to certification. Was this a deliberate attempt by

Boeing to deceive the FAA?; was it just an unfortunate mistake?; or is there some other explanation?

In fact, Boeing engineers did carry out safety assessments on the expanded MCAS. They made a number of questionable assumptions about how pilots would react, in particular, that they would know what to do in an emergency, which later turned out to be quite wrong. On the basis of these assumptions, they appear to have concluded that the extension of MCAS to low-speed situations posed no significant increased risk.[25] That is bad enough. But the main point here is this. Because the re-analysis indicated that there was no appreciable increase in risk, Boeing engineers regarded these analyses as *internal documents* only. FAA guidance does not explicitly specify that documents generated in the process of revision are "required certification deliverables". For this reason, Boeing did not provide any analysis to the FAA. It was still able to claim, however, that it had complied with all relevant FAA guidance. So it was that FAA certification engineers remained in the dark about the expanded capacity of MCAS.[26]

It appears from this account that Boeing engineers were careful to comply with FAA guidance, so far as it went, but not similarly concerned to ensure that the FAA certification engineers were well informed about what Boeing had done. This very literal and limited interpretation of what was required meant that the FAA had no understanding of the real risks that had been introduced by the expansion of MCAS to cover low-speed flight. It is fair to say the FAA engineers were deceived, but it is not clear that this deception was deliberate. The most charitable view is that the problem was Boeing's primary focus on regulatory compliance, rather than safety. On this view, the problem was one of "ritualistic compliance".[27]

A report of the Office of the Inspector General (OIG) highlights this issue. The fundamental problem, it said, was weaknesses in FAA guidance. For example, FAA guidance did not specify that internal revision documents were required certification deliverables, which it should have. It must be said, however, that guidance and procedures can never cover every conceivable situation. Moreover, making changes to guidelines in the U.S. context is enormously difficult because of the politics involved. So, particularly in the U.S., safety depends not only on regulatory compliance but also on good engineering practice and indeed engineering integrity. These more nebulous desiderata were easily overwhelmed by the enormous pressure to get the aircraft certified and off to customers in the shortest possible time. Literal, ritualistic compliance served this purpose admirably.[28]

FAA ORGANISATIONAL SILOS

The Inspector General's report does, however, make another significant observation.[29]

> Communication gaps [within the FAA] further hindered the effectiveness of the certification process. FAA's certification *flight test team* was aware that Boeing had significantly revised MCAS. However, due to a lack of effective coordination within FAA and as well as between FAA and Boeing, some of FAA's *certification engineers* and personnel responsible for approving the level of airline pilot training told us they were unaware of Boeing's changes to MCAS and their impact (emphasis added).

This statement highlights the lack of communication between various sections of the FAA as one of the factors contributing to the certification of the faulty MCAS. The FAA flight test team had taken part, with Boeing, in various flight tests of the 737 MAX. If they had taken a broader view of their responsibilities, this vital information might have been more effectively communicated to the certification team. Their failure to do so amounts to a flaw in FAA structure and functioning.

There is a further inference that can be drawn from the Inspector General's statement. If Boeing had *intended* to keep the FAA in ignorance, it had not been very successful: the crucial information had indeed crossed the line from one organisation to the other, even though it had not ended up where it might have made a difference. This makes the claim of deliberate deception rather less persuasive.

THE SMOKING GUN EMAIL

To return to the chronology, that same month, **November 2016,** the chief technical pilot carried out validation tests of a new MAX *simulator.* He was not test-flying a MAX aircraft but a simulator on the ground. The difference was critical as we shall see in Chapter 8. In his testing, the MCAS fired not just in tight, high-speed turns, as he had told the FAA, but also at low speeds. Moreover, it did so repeatedly, despite his attempt to keep the nose up. In an email to his deputy, he described the plane's *simulated* performance as "egregious" and crazy.[30] His initial thought was that he had misled the FAA – that he had "lied (unknowingly) to the regulator", as he said in the email. That email implies that to this point he had not known about the expansion of MCAS to cover low-speed stalls. It also suggests that despite his new understanding that he had "unknowingly lied", the chief pilot took no action to correct the "lie". A potential explanation for this failure to correct the "lie" will be canvassed in Chapter 8.

The email to his deputy was seen by many as a smoking gun when it came to light in investigations after the crashes. It eventually became the basis for two prosecutions for fraud, one of Boeing itself, and the other of the author of the email, the chief technical pilot, to be discussed in Chapter 8.

In **March 2017**, the 737 MAX was finally certified, and in **May 2017**, distribution to airline customers began.

SUMMARY

The FAA was not resourced to carry out the work necessary to certify a new aircraft as safe. As a result, it had delegated much of this work to specified Boeing employees, which created massive conflicts of interest for these people. The problem was compounded by the failure of the FAA to exercise effective oversight of the Boeing employees to whom it had delegated responsibilities. The system came close to being one of self-certification, which is, of course, predictably ineffective. The result was that the certification process for MCAS was fatally flawed.

Part-way through the process Boeing engineers realised they would need to extend the range of MCAS from the high-speed, tight turns, to the possibility of low-speed stalls. The extension to low speeds was far more dangerous because in this context

MCAS was vulnerable to the single point failure of a very flimsy piece of equipment – an AoA sensor.

Poor communication within Boeing meant that various critical actors in the process were unaware of the change and, hence, of the new risk that had been introduced. Moreover, Boeing engineers had done a safety study of MCAS before its range was extended. After the changes they did a second safety study and convinced themselves that the extension to low speed did not add appreciably to safety risk and that the earlier safety study was still valid. They therefore felt they were within the letter of the law to submit the original safety study to the FAA, making no mention of the changes to MCAS. So it was that the FAA never knew about the changes until after the first MAX accident.

Boeing engineers had a motive to deceive the FAA about these changes. But it is not clear that they *intended* to deceive the regulator. Arguably, less intentional processes provide a better explanation. In particular, the fragmentation of knowledge about MCAS within Boeing meant that no one was accountable for the communications that went to the FAA. Moreover, FAA guidance failed to highlight the things that needed to be highlighted. This uncertainty about what Boeing engineers knew and what they intended became critical in subsequent criminal proceedings, as will now be discussed.

NOTES

1 U.S. House of Representatives, Final Committee Report: Boeing 737 MAX, Executive Summary, p. 24.
2 U.S. House of Representatives, Final Committee Report: Boeing 737 MAX, Executive Summary, p. 19.
3 U.S. Department of Transportation, Office of Inspector General, hereafter OIG, "Weaknesses in FAA's Certification and Delegation Processes Hindered Its Oversight of the 737 MAX 8", February 23, 2021, p. 33.
4 Robison P, *Flying Blind*, Penguin Random House, New York, 2021, pp. 178–179.
5 U.S. Department of Transportation, OIG, "Weaknesses in FAA's Certification and Delegation Processes Hindered Its Oversight of the 737 MAX 8", February 23, 2021, p. 33.
6 U.S. Department of Transportation, OIG, "Weaknesses in FAA's Certification and Delegation Processes Hindered Its Oversight of the 737 MAX 8", February 23, 2021, p. 35.
7 U.S. House of Representatives, Final Committee Report: Boeing 737 MAX, Executive Summary, p. 15.
8 U.S. Department of Transportation, OIG, "Weaknesses in FAA's Certification and Delegation Processes Hindered Its Oversight of the 737 MAX 8", February 23, 2021, p. 31.
9 U.S. Department of Transportation, OIG, "Weaknesses in FAA's Certification and Delegation Processes Hindered Its Oversight of the 737 MAX 8", February 23, 2021, p. 9.
10 U.S. Department of Transportation, OIG, "Weaknesses in FAA's Certification and Delegation Processes Hindered Its Oversight of the 737 MAX 8", February 23, 2021, p. 9.
11 "How the FAA Allows Jet Makers to Self-Certify That Planes Meet U.S. Safety Requirements," *Washington Post*, March 15, 2019.

12 "How the FAA Delegated Oversight to Boeing", *POLITICO*, March 21, 2019.
13 Brief by Moshansky V., presented at the hearings of the House of Commons Standing Committee of Transport, *Infrastructure and Communities*, at Ottawa, ON, on Thursday, April 6, 2017.
14 U.S. House of Representatives, Final Committee Report: Boeing 737 MAX, Executive Summary, p. 15.
15 U.S. House of Representatives, Final Committee Report: Boeing 737 MAX, Executive Summary, p. 20.
16 U.S. House of Representatives, Final Committee Report: Boeing 737 MAX, Executive Summary, p. 21.
17 U.S. House of Representatives, Final Committee Report: Boeing 737 MAX, Executive Summary, p. 16, 24, 25.
18 U.S. House of Representatives, Final Committee Report: Boeing 737 MAX, Executive Summary, p. 25.
19 U.S. House of Representatives, Final Committee Report: Boeing 737 MAX, Executive Summary, p. 22.
20 Robison P, *Flying Blind*, Penguin Random House, New York, 2021, p. 243.
21 "Criminal indictment imminent for former Boeing 737 MAX chief technical pilot, report says", *Seattle Times*, September 17, 2021.
22 See below.
23 U.S. House of Representatives, Final Committee Report: Boeing 737 MAX, Executive Summary, p. 26; Robison P, *Flying Blind*, Penguin Random House, New York, 2021, p. 249.
24 Robison P, *Flying Blind*, Penguin Random House, New York, 2021, p. 249.
25 "Flawed Analysis", *Seattle Times*, March 21, 2019.
26 U.S. Department of Transportation, OIG, "Weaknesses in FAA's Certification and Delegation Processes Hindered Its Oversight of the 737 MAX 8", February 23, 2021, Report, pp. 16–17.
27 Hopkins, *Disastrous Decisions: Human and Organisational Causes of the Gulf of Mexico Blowout*, CCH, 2012, p. 144.
28 There are alternative regulatory regimes that would deal with this problem – safety case regimes. In such regimes the overarching legal requirement is not to comply with particular rules, but rather, to demonstrate to the regulator that risks have been reduced to as low as reasonably practicable. Such a demonstration would require that the risks of MCAS, as installed, be shown to be as low as reasonably practicable. Hopkins, *Disastrous Decisions: Human and Organisational Causes of the Gulf of Mexico Blowout*, CCH, 2012, p. 145ff.
29 U.S. Department of Transportation, OIG, "Weaknesses in FAA's Certification and Delegation Processes Hindered Its Oversight of the 737 MAX 8", February 23, 2021, p. 9.
30 Robison P, *Flying Blind*, Penguin Random House, New York, 2021, p. 251.

8 Exploring the Legal Consequences

The MAX crashes spawned a number of court proceedings which are worth exploring, for various reasons. First, they shed more light on the events leading up to certification by the FAA. Second, they demonstrate some weaknesses and also some of the strengths of law in dealing with corporate disasters. Third, one of these court cases highlights the culpability of the Boeing Board for the two crashes and, in turn, provides an extremely useful set of ideas about what the Boeing Board, and by extension other boards, should do to prevent such accidents. This is an invaluable resource that should be compulsory reading for boards of companies operating in industries at risk of catastrophic accidents.[1]

THE DEPARTMENT OF JUSTICE PROCEEDINGS

In January 2021, nearly two years after the second crash, the Department of Justice (hereafter, DoJ) issued the following statement.

> The Boeing Company (Boeing) has entered into an agreement with the Department of Justice to resolve a criminal charge related to a conspiracy to defraud the [FAA]…
>
> Under the terms of the [agreement], Boeing will pay a total criminal monetary amount of over $2.5 billion…
>
> The tragic crashes of Lion Air Flight 610 and Ethiopian Airlines Flight 302 exposed fraudulent and deceptive conduct by employees of one of the world's leading commercial airplane manufacturers…
>
> Boeing's employees chose the path of profit over candour by concealing material information from the FAA concerning the operation of its 737 MAX airplane and engaging in an effort to cover up their deception. This resolution holds Boeing accountable for its employees' criminal misconduct …
>
> The misleading statements, half-truths, and omissions communicated by Boeing employees to the FAA impeded the government's ability to ensure the safety of the flying public…
>
> As Boeing admitted in court documents, Boeing — through two of its 737 MAX Flight Technical Pilots — deceived the FAA … about an important aircraft part called...MCAS… Because of their deception, a key document published by the FAA lacked information about MCAS, and in turn, airplane manuals and pilot-training materials for U.S.-based airlines lacked information about MCAS.

The statement goes on

> In and around November 2016, two of Boeing's 737 MAX Flight Technical Pilots, one who was then the 737 MAX Chief Technical Pilot …, *discovered information* [emphasis added] about an important change to MCAS. Rather than sharing information about this change with the FAA …, Boeing, through these two 737 MAX Flight Technical Pilots, concealed this information and deceived the FAA about MCAS…

DOI: 10.1201/9781003629269-8

The statement further justified the DoJ's focus on the two technical pilots by observing that

> the misconduct was neither pervasive across the organization, nor undertaken by a large number of employees, nor facilitated by senior management.

Let us consider the agreement in a little more detail. It holds Boeing vicariously liable for criminal fraud allegedly perpetrated by two employees, relatively low in the corporate hierarchy. The role of decision-makers higher in the corporation is ignored. Boeing's admission in the settlement is not that Boeing as a whole, nor its senior officers, did wrong, but that two of its pilots did. They are the sole perpetrators of the "misleading statements, half-truths, and omissions communicated" to the FAA. The DoJ statement in effect presents Boeing as a *victim* of a fraud perpetrated by two employees, for their own personal benefit.

This was problematic for the DoJ. The two 737 MAX crashes with the death of all on board were widely perceived to be a result of Boeing putting profit ahead of safety. This led to widespread demands that Boeing and indeed its CEO and directors should be severely punished. These demands came most stridently from the relatives of people killed. Blaming the two pilots clearly did nothing to satisfy these demands. In fact, the settlement did not even register a conviction against Boeing – that possibility was deferred for three years and depended on Boeing's behaviour during that period.[2] If that were all, Boeing would appear to have gotten off extremely lightly. But that was not all. The settlement acknowledged the community demand for punishment by including a large financial penalty, described as a "criminal monetary amount" of over $2.5 billion. In this way, even though the agreement portrays Boeing as a victim of fraud perpetrated by two employees, the company was nevertheless forced to pay a significant penalty. Leaving aside the adequacy of the "criminal monetary amount", the outcome is, to say the least, bizarre. It seemed to many observers that the two employees had been scapegoated and blamed for what was truly a corporate crime. Presumably there are practical legal reasons for the DoJ proceeding in this way. But it is a very awkward way of dealing with what a lawyer for the families of the people killed described as "the deadliest corporate crime in U.S. history".[3]

The DoJ did not leave the matter there. It subsequently *compounded* the perceived injustice of the outcome by separately prosecuting just the chief technical pilot.

In announcing the new indictment, the Department said:

> In and around November 2016, [the chief technical pilot] discovered information about an important change to MCAS. Rather than sharing information about this change with the FAA..., [he] allegedly intentionally withheld this information and deceived the FAA … about MCAS.[4]

This was essentially the substance of what was said about the two pilots in the indictment of Boeing itself. Note that it was Boeing that had previously admitted that the pilots had committed fraud, not the pilots themselves. However, because the chief pilot was now the target of the prosecution, he defended himself vigorously, bringing new information to light. It turns out that soon after experiencing the "egregious" simulator behaviour, described in Chapter 7; the chief pilot discovered that it was due to a glitch in the *simulator*, which was subsequently fixed. MCAS, as installed in the MAX, was not the source of the problems he experienced in the simulator. He was therefore able to claim, plausibly, that he only came to know about the expanded capability of the MCAS much later, after the Lion Air crash.[5] Additional evidence was given at the trial that, regardless of the simulator behaviour, the chief pilot *did* become aware at this time of the expanded capability of MCAS, but this was contradicted by other evidence. In the end, the charge was not established to the satisfaction of the jury and the pilot was acquitted.[6]

There is a strange irony here. By bringing the prosecution, and losing, the DoJ undermined the entire basis of its earlier agreement with Boeing. If the chief technical pilot was not guilty of fraud, then neither was Boeing. But again, the criminal justice system appears able to live with a truly bizarre outcome.

It is possible that the DoJ's decision to prosecute the chief technical pilot was intended to give expression to a policy announced in 2015, namely,

> It is our obligation at the Justice Department to ensure that we are holding law breakers accountable regardless of whether they commit their crimes on the street corner or in the boardroom. In the white-collar context, that means pursuing not just corporate entities, *but also the individuals through which these corporations act.* [emphasis added].[7]

That is an admirable policy position, but if this was the basis for singling out the pilot for prosecution, it was a misapplication of the policy. The fact is there were individual members of the Board who might more appropriately have been held criminally liable. I say this because in another legal matter arising out of the MAX crashes, to be discussed shortly, the judge concluded that the Lead Director of the Boeing Board had knowingly made false public statements, by claiming that the Board had been diligent in overseeing the safety of the MAX, when in fact it had not.

Even if fraud by two pilots had been established convincingly, we would still be a long way from understanding what happened. The concept of fraud fails to embrace the complexity of these events and is arguably a hindrance to any understanding. Moreover, the rules of criminal procedure are largely aimed at protecting the rights of defendants, not necessarily getting at the truth. Civil actions for damages are therefore often more enlightening.

THE SHAREHOLDER JUDGEMENT

I turn now to a civil judgement in the Boeing 737 matter, handed down in September 2021, which provides much greater understanding and, in turn, more important insights into prevention, than did the DoJ proceedings.[8]

The case was brought by shareholders against the Boeing company directors and sought to show that the directors had failed them in relation to airplane safety. The judgement (hereafter the shareholder judgement) is scathing. It found that the directors had

i. before the Lion Air crash, utterly failed to establish a system for monitoring and overseeing airplane safety, and
ii. after the Lion Air crash, when confronted with "red flags" about the safety of MCAS", consciously disregarded their duty to investigate.[9]

The judgement provides a clear indication of how the Board should have behaved and for that reason is worth considering in more detail.

The court described aircraft safety as a "mission-critical" aspect of the company's business, which required the Board to exercise its oversight function rigorously, something it had failed to do. (All quotes below are from the judgement, unless the context indicates otherwise.)

The concept of "mission-critical" came from another case on which the court drew as a precedent. For present purposes, it can be equated with the concept of *major accident risk* used in hazardous industries such as petroleum, mining, power generation, and chemical manufacturing. Major accident risks must be managed differently from the way more conventional health and safety workplace hazards are managed. They require special attention, including by company boards. It is possible for a company to appear to do well in managing conventional workforce health and safety and yet be managing its major accident risks badly. That was one of the findings of inquiries into the BP Texas City refinery fire of 2005 and the BP oil well blowout in the Gulf of Mexico in 2010, each of which killed numerous workers.[10] Similarly, Boeing's mission-critical safety risks needed to be managed differently. In October 2018, just seven days before the Lion Air crash, the company won a prestigious award from the U.S. National Safety Council (NSC).[11] "Boeing is a leader in one of the most safety-centric industries in the world", the NSC president said. "Ergonomic issues are a common risk in aircraft manufacturing environments. Boeing engineers and employees have worked together to create specialised tools for at-risk jobs", she said. But regardless of how well Boeing may have been managing these workplace risks, at the very same time it was badly mismanaging the risk of aircraft safety.

So how *did* the Boeing Board deal with aircraft safety? The short answer is: it didn't. The shareholder judgement identifies a number of specific ways in which the Board lacked focus on aircraft safety. In so doing, it points to ways in which this mission-critical risk could be better managed. I deal with several here.

For a start, "the board had no sub-committee charged with direct responsibility to monitor airplane safety". It established such a committee only after the two crashes.[12]

Boeing did have an audit committee charged with "risk oversight" but the audit committee did not see risk to safety as part of its remit. It was concerned primarily with financial risks. The audit committee's yearly report on compliance risk management did not include airplane safety. This kind of risk-blindness is remarkably widespread in the corporate world. For senior executives and boards of big companies, even those operating in hazardous industries, risk management almost always refers to financial risk management. Even where safety regulators require companies to do risk assessments, with the expectation that these are safety risk assessments, that expectation is often lost sight of by people doing the risk assessment.[13]

Furthermore, there was no mechanism for whistleblowers and employees to bring their safety concerns to the Board's attention.[14] While some of these complaints made their way to senior managers, none made it to the Board. In particular, the Board was unaware of complaints about aircraft safety, compliance, workforce exhaustion, and production schedule pressures.[15]

Nor was there any mechanism for bringing to the Board's attention safety issues (red flags) of which senior Boeing management was well aware.[16]

All these matters demonstrate that the Board failed to implement any system to inform itself on the company's "mission-critical" safety performance.[17]

More generally, aircraft safety was not a routine item on the Board agenda.[18] After the Lion Air crash in October 2018, the crash did not appear on the formal agenda until the regularly scheduled December meeting. Even then the discussion concerned restoration of profitability and efficiency – not product safety, MCAS or the AoA sensor. The Board did allocate five minutes to the discussion of a legal memo "including matters related to the Lion Air incident",[19] but that was the only context in which safety might have been mentioned. It was not until April 2019, after the FAA grounded the 737 MAX fleet, that the Board devoted time to a discussion of safety.

The Board claimed in its defence that "it oversaw the quality and safety of the 737 MAX through monitoring the progress of the FAA's extensive certification review of the 737 MAX".[20] The judge was dismissive of this claim. As he said

> the Board focussed on the 737 MAX's production, development and certification in order to assess production timelines and revenue expectations, and to strengthen the Company's relationships with FAA officials -not to consider customer safety.[21]

In one of his most important remarks, he goes on to say,

> the fact that the company's product [on the face of it] satisfied regulatory requirements does not mean that the board has fulfilled its oversight obligations to prevent corporate trauma.[22]

All boards should take this comment to heart. Time and again in accident investigations, one comes across cases where the failure of the regulatory approval process has been one of the factors contributing to an accident. This is such a critical point that I give two more examples.

Consider the blowout on the drilling rig, Deepwater Horizon, in the Gulf of Mexico in 2010.[23] The exploration drilling operation was complete; a vast reservoir of oil and gas, under very high pressure, had been located 13,000 feet below the seabed; the rig needed to pump cement into the bottom of the exploratory well to plug it, before moving on to its next assignment. The regulatory requirement was that the cement plug in the well extends to 500 feet above the level where oil and gas had been found, as a margin for safety, in case the cement job was in any way defective (which it later turned out to be). BP's local well engineers wanted to cut that margin to 100 feet. The question they asked themselves was not: "is this an adequate safety margin?" but "can we get regulatory approval for it?" The regulator generally approved such requests, and often within a matter of hours, relying on the information provided by the company. The result of this practice was that the company could claim it had regulatory approval for what it wanted to do, based on a regulatory decision that was probably beyond the competence of the regulator. In the Deepwater Horizon case, there was only one person in the regulator's regional office performing this function. He approved the request. It is not clear to what extent this particular decision contributed to the blowout, but it was part of a pattern of decision-making that focussed on compliance, not safety – a pattern that certainly contributed to the accident.

Here is another documented example of the problem.[24] In January 2019, the mining company Vale suffered the collapse of one of its tailings dams in Brazil, killing 270 people. One of the Vale directors said at the time the Vale Board had no idea how badly risk was being managed.[25] A report provided to the Board less than six months earlier said this:

> 100% of Vale iron ore dams were audited in August 2018 and had a stability declaration issued by the external auditor with certified safety conditions. All dams are safe, stable and operate within normal range.

How, then, could one of these dams collapse? The fact is that pressure was brought to bear on auditors to give favourable reports on these dams and the dam in question should never have been certified safe.

The director said later she had learnt her lesson, which was to be sceptical of the routine good news reports the Board received and to probe for the bad news – "challenge the green and embrace the red", as it is often put. This is a mindset that should be assiduously cultivated by boards of companies in industries where catastrophic accidents can occur.

The Boeing Board assumed, without question, that certification meant any safety risks had been dealt with and there was no need for any scepticism about the certification process. The court judgement highlights the fact that boards can be held to be at fault if they fail to challenge the green and embrace the red. All boards should take this to heart.

One conspicuous way in which the Boeing Board failed to challenge the green was its failure to challenge any of the CEO's statements about safety, particularly after the first crash. It did not ask him for a briefing on that accident and waited till he saw fit to raise the matter, which he did in an email one week after the crash. In that email, he treated the crash as a public relations problem and told the Board the "737 MAX fleet is safe". His main communications to the board, in the weeks following, focussed on discrediting media reports about the MCAS and blaming the Lion Air repair shops and crew.[26] His claim that the aircraft was safe was based on what he described as the "rigorous test program" Boeing had undergone to earn FAA certification. This flew in the face of mounting evidence that newspapers were reporting about the problems with MCAS and the AoA sensor.[27] The CEO's claim highlights the utter foolishness of relying on regulatory compliance, in particular certification, as the sole assurance of safety.

Even after the second crash, "the Board passively accepted… [the CEO's] assurances … that the public and regulatory backlash [to the grounding by the FAA], was driven solely by public/political pressure, not by any new facts about the 737 MAX's safety. The Board did not press for more information".[28] One Board member even praised the CEO for his "strong leadership" in engaging with top government officials in an effort to keep the MAX flying.[29] It is hard to imagine a board *less* concerned with safety.

In the preceding discussion, I have emphasised aspects of the judgement that are most helpful in understanding how the Board failed in its duty to oversee aircraft safety. That emphasis allows us to see more clearly what boards in general need to do to avoid similar disasters.

But the case was not just about identifying lessons, it was about assigning culpability. The degree of culpability depends crucially on whether the offender *knowingly* offended. The judge found that to be the case, as follows:

> The Board publicly lied about if and how it monitored the 737 MAX's safety. The Lead Director ... held a series of interviews with major newspapers with the following corporate objective: "Position the Boeing Board of Directors as an independent body that has exercised appropriate oversight". As to the Lion Air crash, he represented that the Board had been "notified immediately, as a board broadly", after the Lion Air crash and met "very, very quickly" thereafter, participated in evaluating the safety risk associated with the 737 MAX; and considered grounding the 737 MAX after the Lion Air crash, but concluded the crash "was an anomaly" that did not warrant grounding the airplane. As to the Ethiopian Airline crash he represented that the Board met within twenty-four hours of the crash to discuss potential grounding of the 737 MAX and recommended that the 737 MAX be grounded. Each of these representations was false.[30]

The judge emphasised that the Director's statements were *knowingly* false and furthermore, that he "knew what the Board should have been doing all along".[31] In the judge's view, the Board was highly culpable.

There is another inference we can draw from the shareholder judgement. The Board's complete failure to attend to aircraft safety, coupled with its concern to maximise production, was a much more significant causal factor than the alleged fraud of the two technical pilots. It is a great pity that the DoJ was not able to raise its gaze to Board level in the way that the civil court did.

As a result of the civil court's decision, the shareholders reached a settlement agreement with Boeing that involved payment to shareholders of $238 million. More importantly for safety, the agreement included a commitment from Boeing that it would make a number of organisational changes.[32] These included the following promises:

- At least three Board directors would have knowledge, experience, and/or expertise with aviation/aerospace, engineering, and/or product safety oversight.
- Top management of the company would include a chief engineer and a chief aerospace safety officer.
- The board would have an Aerospace Safety Committee (ASC) consisting of three or more independent directors (i.e., not including the CEO) with the relevant safety expertise.
- The chief engineer and a chief aerospace officer would report to the ASC at least twice a year on a number of quite specific safety issues, including the Boeing whistleblower channel – the Speak Up portal – and the Committee would be provided with any original documentation it wanted on these issues.
- Bonuses paid to top (named) executive officers would include consideration of safety metrics. These would include metrics "best suited to encourage strong oversight of safety improvements and overall safety of company products".[33]
- The company would appoint an ombudsperson, reporting to the chief aerospace safety officer, providing, amongst other things, an avenue for FAA-authorised representatives to raise concerns or make complaints.

BOEING'S ORGANISATIONAL RESPONSE

The shareholder settlement was finalised in November 2021, nearly three years after the Ethiopian Airlines crash. But Boeing had not been idle in the interim. Shortly after the grounding of the 737 MAX, a small Committee of the Board began interviewing a range of safety experts and senior executives in high-hazard industries to determine how Boeing could re-organise itself to make safer aircraft. Five months later, they delivered their findings to the Board, which duly accepted them.[34] Here are some of their recommendations.

There should be a new aerospace safety sub-committee of the Board, to be chaired by a retired admiral, a career nuclear trained submarine officer. This was an inspired choice. The U.S. nuclear navy has the reputation of being one of the most safety-conscious organisations in the world.[35] After the Columbia space shuttle disaster in 2003, the Columbia Accident Investigation Board of Inquiry looked to the U.S. nuclear navy for ideas about how NASA might re-organise itself to give safety a higher priority.[36]

One of the committee's most significant recommendations was aimed at empowering engineers, by restructuring their reporting lines. Under existing arrangements, engineers reported primarily to the business leaders for each airplane model and only in a secondary and nebulous way to more senior engineers. Under the new model, they would primarily report through to a chief engineer who in turn reported to the CEO.[37] There would still be a secondary reporting line to the business managers, but performance evaluations would depend primarily on more senior engineers, not business managers. This was an organisational design, drawn from the nuclear navy, that the Columbia Accident Investigation Board had recommended for NASA. It is also the organisational design that BP adopted after the Deepwater Horizon accident, which cost BP upwards of $65 billion.[38] The Goldman Sachs Group and Volkswagen have undertaken similar reforms in recent settlements with prosecutors.[39] It seems that only after major incidents with severe financial impacts do companies give serious consideration to such organisational restructuring.

Unfortunately, a Congressional review more than four years later found that this organisational re-structuring at Boeing had still not occurred.[40]

A further recommendation from the Boeing Board was that a new Product and Services Safety organisation be created, reporting directly to senior company leadership and the Board's Aerospace Safety Committee. Among other things, Boeing's engineering and technical experts who represent the FAA Administration in airplane certification activities would report through this structure.

It is clear from this account that there is a strong overlap between the decisions the Board came to (five months after the grounding of the MAX) and the agreement with shareholders (made two years later) in relation to organisational changes that Boeing needed to make. No doubt the former influenced the latter. But the beauty of the shareholder judgement is that it enshrines these changes in a legal agreement which will give them a much greater degree of permanence and a mechanism by which shareholders will more easily be able to hold the Board to account. The other benefit of the shareholder agreement is that it is accompanied by a detailed judgement that provides a great deal of evidence about what went wrong and why.

Only time will tell how effectively Boeing implements all the elements of the shareholder agreement, or indeed the organisational changes specified by the Board five months after the grounding of the MAX. It is to be hoped that the external compliance monitor appointed under the separate 2024 DoJ agreement[41] will take an interest in this matter.

NOTES

1 Discussion of the Aircraft Certification, Safety and Accountability Act 2020 will be deferred to Chapter 10.

2 In fact, the DoJ took the view that the door plug blowout in January 2024 violated the terms of the agreement. This led Boeing to agree in July 2024 to plead guilty to a felony charge of conspiring to defraud the federal government in relation to the original MCAS certification. "Boeing Agrees to Plead Guilty to Felony Deal with Justice Department", *New York Times*, August 7, 2024.

3 "Relatives of people killed in 2 Boeing MAX crashes ask the US to fine the company $24.8 billion," *Seattle Times* June 19, 2024. https://www.seattletimes.com/business/relatives-of-those-killed-in-boeing-max-crashes-ask-u-s-to-fine-the-company-24-8b/

4 "Former Boeing 737 MAX Chief Technical Pilot Indicted for Fraud", Archives US DoJ, October 14, 2024. https://www.justice.gov/opa/pr/former-boeing-737-max-chief-technical-pilot-indicted-fraud

5 "Why Former Boeing 737 MAX Chief Technical Pilot Indicted for Fraud. Boeing Pilot Forkner Was Acquitted in the 737 MAX Prosecution", *Seattle Times*, March 25, 2022.

6 "Why Boeing Pilot Forkner Was Acquitted in the 737 MAX Prosecution", *Seattle Times*, March 25, 2022.

7 The so-called Yates memo. See Avergun J., et al., "Corporations, Directors, and Officers: Potential Criminal and Civil Liability. A Lexis Practice Advisor Practice Note", p. 5.

8 IN RE THE BOEING COMPANY DERIVATIVE LITIGATION, Court of Chancery of Delaware, September 7, 2021, https://casetext.com/case/in-re-the-boeing-co-derivative-litig. Hereafter Shareholder judgment. For legal commentary see https://corpgov.law.harvard.edu/2022/06/01/revisiting-the-boards-oversight-role-after-in-re-boeing-co/

9 IN RE THE BOEING COMPANY DERIVATIVE LITIGATION, Court of Chancery of Delaware, September 7, 2021, https://casetext.com/case/in-re-the-boeing-co-derivative-litig, pp. 59, 70–71.

10 See details and references in my books, *Failure to Learn: the BP Texas City Refinery Disaster*, CCH, 2008; *Disastrous Decisions: Human and Organisational Causes of the Gulf of Mexico Blowout*, CCH, 2012.

11 https://www.nsc.org/in-the-newsroom/nsc-presents-the-boeing-company-with-2018-robert-w-campbell-award

12 IN RE THE BOEING COMPANY DERIVATIVE LITIGATION, Court of Chancery of Delaware, September 7, 2021, https://casetext.com/case/in-re-the-boeing-co-derivative-litig, p. 74.

13 See Hopkins A, *Sacrificing Safely: Lessons for Chief Executives*, Wolters Kluwer, Sydney, 2022, pp. 43–45.

14 IN RE THE BOEING COMPANY DERIVATIVE LITIGATION, Court of Chancery of Delaware, September 7, 2021, https://casetext.com/case/in-re-the-boeing-co-derivative-litig, p. 76.

15 IN RE THE BOEING COMPANY DERIVATIVE LITIGATION, Court of Chancery of Delaware, September 7, 2021, https://casetext.com/case/in-re-the-boeing-co-derivative-litig, p. 32.

16 IN RE THE BOEING COMPANY DERIVATIVE LITIGATION, Court of Chancery of Delaware, September 7, 2021, https://casetext.com/case/in-re-the-boeing-co-derivative-litig, p. 87.
17 IN RE THE BOEING COMPANY DERIVATIVE LITIGATION, Court of Chancery of Delaware, September 7, 2021, https://casetext.com/case/in-re-the-boeing-co-derivative-litig, p. 89.
18 IN RE THE BOEING COMPANY DERIVATIVE LITIGATION, Court of Chancery of Delaware, September 7, 2021, https://casetext.com/case/in-re-the-boeing-co-derivative-litig, p. 76.
19 IN RE THE BOEING COMPANY DERIVATIVE LITIGATION, Court of Chancery of Delaware, September 7, 2021, https://casetext.com/case/in-re-the-boeing-co-derivative-litig, pp. 39–40.
20 IN RE THE BOEING COMPANY DERIVATIVE LITIGATION, Court of Chancery of Delaware, September 7, 2021, https://casetext.com/case/in-re-the-boeing-co-derivative-litig, p. 78.
21 IN RE THE BOEING COMPANY DERIVATIVE LITIGATION, Court of Chancery of Delaware, September 7, 2021, https://casetext.com/case/in-re-the-boeing-co-derivative-litig, pp. 78–79.
22 IN RE THE BOEING COMPANY DERIVATIVE LITIGATION, Court of Chancery of Delaware, September 7, 2021, https://casetext.com/case/in-re-the-boeing-co-derivative-litig, p. 80.
23 See Hopkins, *Disastrous Decisions: Human and Organisational Causes of the Gulf of Mexico Blowout*, CCH, 2012, Chapter 10.
24 See Hopkins A and Kemp D, *Credibility Crisis: Brumadinho and the Politics of Mining Industry Reform,* Walters Kluwer, Sydney, 2021.
25 Interview with Guerra S, *The Hawkamah Journal*, Issue 16, December 2020.
26 IN RE THE BOEING COMPANY DERIVATIVE LITIGATION, Court of Chancery of Delaware, September 7, 2021, https://casetext.com/case/in-re-the-boeing-co-derivative-litig, p. 82.
27 IN RE THE BOEING COMPANY DERIVATIVE LITIGATION, Court of Chancery of Delaware, September 7, 2021, https://casetext.com/case/in-re-the-boeing-co-derivative-litig, pp. 82–83.
28 IN RE THE BOEING COMPANY DERIVATIVE LITIGATION, Court of Chancery of Delaware, September 7, 2021, https://casetext.com/case/in-re-the-boeing-co-derivative-litig, p. 84.
29 IN RE THE BOEING COMPANY DERIVATIVE LITIGATION, Court of Chancery of Delaware, September 7, 2021, https://casetext.com/case/in-re-the-boeing-co-derivative-litig, pp. 84–85.
30 IN RE THE BOEING COMPANY DERIVATIVE LITIGATION, Court of Chancery of Delaware, September 7, 2021, https://casetext.com/case/in-re-the-boeing-co-derivative-litig, pp. 55–56. This text is reproduced from the "verified amended consolidated complaint", p. 2 of judgment. The actual words of the judge, contained in the "analysis" section of the judgment are a slight abridgement of the words used here. But the abridgement is garbled at points, so is not quoted directly here. Interested readers will find the judge's actual words at p. 91.
31 IN RE THE BOEING COMPANY DERIVATIVE LITIGATION, Court of Chancery of Delaware, September 7, 2021, https://casetext.com/case/in-re-the-boeing-co-derivative-litig, p. 91.
32 In the Court of Chancery of the State of Delaware: In Re The Boeing Company Derivative Litigation – Consol. C.A. No. 2019-0907-Mtz Stipulation and Agreement of Compromise, Settlement, and Release. https://www.osc.ny.gov/files/press/pdf/stipulation-of-settlement-boeing.pdf, Exhibit A.

33 In the Court of Chancery of the State of Delaware: In Re The Boeing Company Derivative Litigation – Consol. C.A. No. 2019-0907-Mtz Stipulation and Agreement of Compromise, Settlement, and Release. https://www.osc.ny.gov/files/press/pdf/stipulation-of-settlement-boeing.pdf, Exhibit A, p. 3.
34 https://boeing.mediaroom.com/2019-09-25-Boeing-Chairman-President-and-CEO-Dennis-Muilenburg-and-Boeing-Board-of-Directors-Reaffirm-Companys-Commitment-to-Safety
35 See Digeronimo M and Koonce B, *Extreme Operational Excellence, Applying the U.S. Nuclear Submarine Culture to Your Organization*. Outskirts Press, 2016.
36 See references in Hopkins, *Organising for Safety: How Structure Creates Culture*, Wolters Kluwer, Sydney, 2019.
37 Boeing Board to Call for Safety Changes after 737 Max Crashes, *New York Times,* September 15, 2019. https://www.nytimes.com/2019/09/15/business/boeing-safety-737-max.html
38 Hopkins, *Organising for Safety: How Structure Creates Culture*, Wolters Kluwer, Sydney, 2019.
39 "Prosecutors Credited Boeing for Compliance, Organizational Reforms", *Wall Street Journal*, January 15, 2021. https://www.wsj.com/articles/prosecutors-credited-boeing-for-compliance-organizational-reforms-11610723642?mod=djemRiskCompliance
40 Expert Panel Review Report, FAA, February 26, 2024, p. 37. https://www.faa.gov/regulationspolicies/rulemaking/committees/documents/section-103-organization-designation
41 "Boeing Struck a Deal with the Justice Department. What Comes Next?" *New York Times*, August 7, 2024; Boeing Agrees to Plead Guilty to Felony Deal with Justice Department, *New York Times*, August 7, 2024.

9 An Evolving Regime of Incentives

Boeing's public statements frequently claim that safety is its top priority. Yet its bonus arrangements demonstrate that in the past this has been far from true. For years, the financial rewards paid to top executives for commercial success have far outweighed any incentives provided for managing safety diligently.[1] The contradiction has been so glaring that those who claim that safety is the top priority must have been blinded by the glare.

Major accidents in high-hazard industries can often be traced back to these financial imperatives. Chapter 11 will deal more systematically with this question of causation, but for present purposes, it is useful to see financial incentives as the root cause of these events. Accident prevention depends on neutralising or counteracting the effect of these incentives. Unfortunately, too few accident investigations highlight this cause. Boeing itself now recognises this causal connection, as demonstrated by changes to its remuneration system for senior executives in 2024.

This chapter begins with an examination of the incentive arrangements for top Boeing executives in place in 2022, three years after the second crash. We shall see that, even at this time, these arrangements continued to relegate safety to a very subordinate position. Thereafter, the system began to evolve, particularly after the Alaska Airlines' door plug accident of January 2024. History may judge this to have been a significant turning point for Boeing, more significant even than the two crashes.

COMPONENTS OF REMUNERATION

There are three components that go into payments made to Boeing's top executives:

- Base salary
- Earned annual incentive awards
- Long-term incentive awards

The first is a fixed amount, while the second and third are variable, "at-risk" components. Figure 9.1[2] shows the total payment made to the CEO in 2022. The commonly used term in North America in this context is "compensation". In other countries, "compensation" is reserved for payments made for losses of various types, for example, payments made after an accident. I use the term remuneration here, to avoid that confusion. The figure shows that the CEO's salary, though considerable – $1.4 million – was a tiny fraction of his total remuneration – only 6%. His long-term incentive (LTI) awards formed the largest fraction – 78%, and his annual incentive payment amounted to 16% of the total.

DOI: 10.1201/9781003629269-9

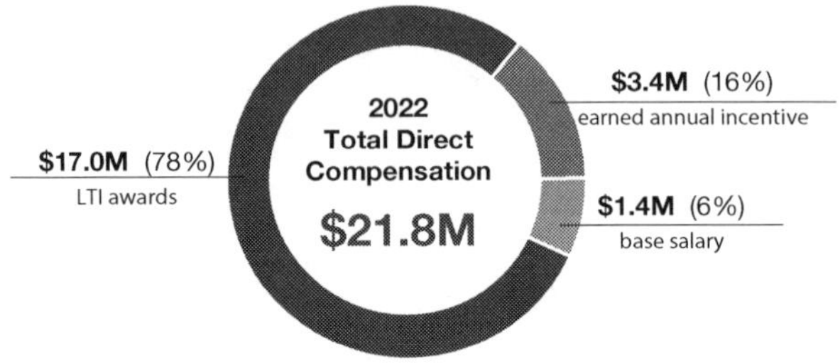

FIGURE 9.1 Components of the CEO's remuneration in 2022.

There are five other top executive officers, for whom these percentage figures are roughly the same, although the total dollar amounts are less than for the CEO. In all these cases, then, the long-term incentives far outweigh the other components of remuneration. I shall focus here on the CEO.

Long-Term Incentives

As explained in Chapter 4, long-term incentives are designed to ensure that, as far as possible, the interest of executives coincides with the interests of shareholders. The awards are made each year, but they are deferred for a period of three years and the actual amount paid depends in various ways on the company's financial performance during that three-year period. The aim is to keep the eyes of executives firmly focussed on the future profitability of the company.

Safety had no impact on the way long-term incentives were determined in 2022. Of course, in the event of a major accident the share price may suffer as indeed it has since the MAX crashes,[3] and long-term awards will then be curtailed, but recall that even after the first 737 crash, Boeing's share price continued on upwards to a record high. It was only after the grounding of the fleets around the world that the share price fell. Accidents of sufficient magnitude to have a significant effect on the share price of a large company are rare for any one company and are very unlikely to occur during the term of office of any particular senior executive. Hence, they are not likely to loom large in executive thinking about how best to maximise their remuneration on an annual basis.

There is one slight qualification that needs to be noted. In 2022, for the first time, safety made an appearance in Boeing's long-term incentive system. In order to qualify for an award, the executive had to complete the company's Safety Management System training. However, this did nothing to change the structure of long-term incentives, so its effect could be expected to be minimal.[4]

Annual Incentive Payment

The second most important component in the total remuneration package is the annual incentive payment. The basic formula for this is presented in Figure 9.2, taken from Boeing's annual statement prepared for shareholders.[5]

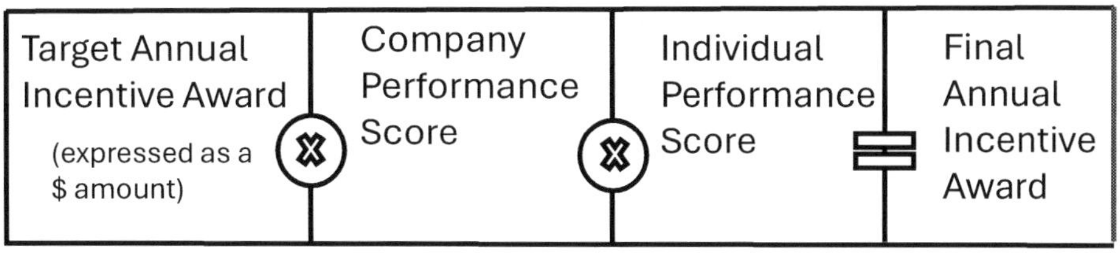

FIGURE 9.2 Components of annual incentive payment for senior executive officers in 2022.

There are several aspects of this formula that need to be understood before we can form a view of the extent to which safety contributed to the CEO's annual incentive payment.

Let us first examine the makeup of the company performance score (second box in Figure 9.2). In the years prior to the MAX crashes, company performance scores were entirely based on financial performance. By 2022, this had evolved to the formula shown in Figure 9.3.[6]

The company performance score now had two components: financial performance which was given a 75% weighting and operational performance which had a 25% weighting, as shown in Figure 9.3. Safety was included in the operational performance. There were five components of operational performance:

- Product safety (20%)
- Employee safety (20%)
- Quality (20%)
- Climate (20%)
- Equity, diversity, and inclusion (20%).

Each of these is weighted equally at 20%. For present purposes, let us focus on airplane safety. Attention to both product safety and quality is likely to improve aircraft safety. So let us say that aircraft safety contributes to 40% of the operational performance measure. This amounts to 10% of the total company performance score (40% of 25%). So, to restate the situation, financial performance contributes 75%, while airplane safety contributes only 10% to the total company performance score.

To understand boxes 1 and 3 in Figure 9.2, we must focus on an individual executive of interest, in this case, the CEO.[7] The first box – the target annual award – is set ultimately by the Board and takes account of the CEO's "experience", "leadership",

FIGURE 9.3 Components of the company performance score in 2022.

and "performance" and the need to ensure his pay is "market-competitive". Clearly, assembling these factors into a single-dollar target is a subjective, and rather arbitrary process. But more importantly, for present purposes, airplane safety is not an explicit factor and, as we know from the shareholder settlement,[8] it did not figure at all in the Board's thinking prior to the 737 crashes.

The third box in Figure 9.2, the individual performance score, is again subjectively assessed, but in the case of the CEO in 2022, the score took into consideration accomplishments in six areas, three of which were related to product safety.[9] These were

- creation of a record number of engineering design practices (procedures for engineering design)
- introduction of a safety management system (as required by a Congressional Act of 2020)
- championing Boeing's policy of Seek, Speak, and Listen.

Potentially this might have resulted in a considerable emphasis on aircraft safety.

However, the final individual performance score settled on by the Board for the CEO in 2022 was 110%. The important thing to notice about this is that multiplying by 110% (i.e., by 1.10) has little impact on the final annual incentive payment.

To summarise the CEO's annual incentive for 2022 was largely a function of two factors – an initial target, determined in a highly subjective way by the Board with no explicit reference to aircraft safety, and a company performance score which overwhelmingly emphasised financial performance while giving very little weight to aircraft safety.

Combining Annual and Long-Term Bonuses

If we now put together the two at-risk components of the CEO's remuneration – long-term bonus and annual bonus – we conclude that the incentives for maximising financial performance dwarfed any incentives to focus on aircraft safety. This situation is completely at odds with the claim that safety is Boeing's top priority.

CHANGES AFTER THE DOOR PLUG BLOWOUT

In early 2024, following the Alaska Airlines door plug blowout, the incentive scheme changed dramatically. Recall that the *company* performance score discussed above was weighted 75% towards financial performance and 25% towards operational performance. After the Alaskan Airlines incident, this imbalance was effectively reversed for the commercial airplane business: 40% for financial performance and 60% for operational performance.[10] Moreover, the operational performance, previously consisting of five components, would now focus entirely on quality and safety targets – again a significant change. (The KPIs – Key Performance Indicators – used for these quality and safety targets will be discussed shortly.) This amounted to a very significant refocussing of the company performance score which did indeed put aircraft safety ahead of financial considerations. The company performance score is the critical determinant of bonuses for employees not eligible for long-term incentive payments. So this change could be expected to have a real impact on managers and workers assembling aircraft on the factory floor. Time will tell.

The situation was different for top executives eligible for long-term incentive payments.[11] The total value of their remuneration would still be primarily dependent on company financial performance. But that was now significantly qualified. A substantial portion of this long-term payment (55%) would be subject to a "product safety downward modifier". Avoiding this downward modifier would depend on the company satisfying two conditions:

- design and deployment of an employee culture survey
- development of operational control on assembly lines.

Failure to achieve these goals within a one-year time frame would result in a reduction by 25% in the relevant long-term incentive payment, and failure to achieve them within two years would result in the cancellation of the payment.

It is not at all clear how an employee culture survey can improve plane safety, as will be discussed in Chapter 10. However, the second of the two conditions above had the potential to have a big impact, as we shall see in a moment. It would mean that top management had a strong financial interest in ensuring that the company met some quite specific plane safety requirements. Note that this is not something that they could personally achieve, unlike completing a safety management training course themselves. These were truly corporate goals, which means that top executives would need to do what they could to facilitate company progress towards these goals. As far as can be ascertained, this is the first time such a safety objective has been included in long-term incentive arrangements for senior Boeing executives. This is potentially a hugely important development. But again, only time will tell.

SAFETY INDICATORS

So far, I have left the product or plane safety goals vague. Let us consider in more detail the plane safety indicators that were to be included in both the corporate performance score and long-term incentive calculations, in 2024. This is another matter in which Boeing appears to have made a real advance.

I begin with a detour. If we are looking for an indicator of how safe a manufacturing workplace is for workers, then injury rate is a reasonable place to start: the lower the injury rate, the safer the workplace can be assumed to be. It is true that injury rates can be manipulated to present an employer in a favourable light, but let us ignore this for present purposes.[12]

The number of injuries is, potentially, a reasonable indicator of workplace safety where injuries occur sufficiently frequently to be able to talk meaningfully about a rate, which can be monitored over time, providing evidence of changing levels of risk. The effectiveness of a new risk reduction strategy can then be assessed by its impact over time on the injury rate.

In the case of major accidents in hazardous industries, such as oil well blowouts, tailings dam failures, coal mine explosions, train crashes, and aircraft crashes, these events are too rare for us to compute frequency rates that can serve as indicators of risk in particular contexts. However, the important feature of such accidents is that they are invariably preceded by warning signs, or precursor events, that normally do not culminate in a major accident, but which indicate an elevated risk of such an accident. If these precursor events occur frequently enough, their rate of occurrence can be treated as an indicator of risk, to be monitored and, if necessary, responded to. Moreover, the most effective major accident prevention strategy is not to wait for a major accident and then seek to learn from it, but rather to work at minimising the number of precursor events. This will become clear if we consider some examples.

- For air traffic control organisations, the worst imaginable accident is a mid-air collision, which fortunately is very rare. The most effective way to reduce the risk of such an accident is to reduce the frequency of the precursor events. The relevant precursor event is the failure of two aircraft to maintain the degree of separation required by air traffic control. In some countries, these are referred to as breakdowns of separation; in the U.S. they are called operational errors.[13] These events occur far more frequently than mid-air collisions themselves. But by monitoring and seeking to minimise the frequency of such events, air traffic control organisations are able to minimise the risk of mid-air collision.
- For petrochemical industries, explosions and major fires are preceded by LOCs – Losses of Containment of flammable substances from the pipes or tanks in which they are supposed to be contained. LOCs are occasions of heightened risk, even though in many cases they do not result in an ignition. But the industry, both the regulators and the regulated, treats the number of LOCs as a critical indicator of how well the risk of major accidents is being managed.
- For underground coal mines, where one of the worst imaginable accidents is a coal gas explosion, a suitable precursor event is the presence of a high concentration of explosive gas, often called an exceedance.
- Train crashes are often preceded by trains failing to stop at a red signal, a so-called signal passed at danger, or SPAD, in the U.K., or SSO, stop signal overrun, in the U.S. Clearly by reducing the frequency of such events, the risk of train crash is reduced.

Evidently, useful indicators of major accident risk are highly context specific and vary with industry, even enterprise, and with the major accident envisaged. Moreover, it may require some ingenuity or research to identify suitable indicators.

Let us consider how this kind of thinking might have prevented the Alaska Airlines incident – the blowout of a door-sized panel in the side of a 737 MAX a few minutes after take-off.

Months before the accident, the plane concerned spent nearly three weeks shuffling down an assembly line with some faulty rivets in need of repair. Workers had spotted the problem almost immediately after the plane's fuselage arrived at the factory for final assembly. But they didn't make the fix right away, and the aircraft continued on to successive workstations. Workers completed the repair 19 days later, after the aircraft had left the assembly line. At this time, they accidentally failed to replace four critical bolts on the door plug that they had removed to do the job, resulting in the door plug blowout a few weeks later.

The incident triggered a six-week audit of the production of the 737 MAX by the Federal Aviation Administration. The audit found dozens of problems throughout the manufacturing process at Boeing, and also at one of its suppliers, Spirit AeroSystems, that supplied the 737 fuselages for assembly in Seattle. The problems all involved failure to comply with quality-control requirements.[14]

Boeing acknowledged that the Alaska Airlines door plug blowout was a result of poor-quality work that was occurring on the aircraft assembly line. It further recognised that this poor-quality work was a result of commercial pressure to keep the line moving as fast as possible to deliver the target number of new aircraft each month. This could result in what Boeing calls "travelled work",[15] as follows.

Each plane moves through a series of stations, where crews must complete certain tasks. Those stations are equipped with tooling and platforms, and with a crew trained to do the jobs designated for the site. Planes advance to the next station roughly every 24 hours.

Sometimes, a missing part prevents workers from finishing the designated job. Leaving the plane sitting in place would slow the entire production line. So the plane moves ahead and the part gets added or the repair is completed somewhere down the line. This is called "travelled work". In some cases, the work isn't done until the plane leaves the factory, to be positioned on the flight line, a spot outside the factory where planes are parked as they await delivery.

Doing a job away from the intended workstation can be problematic. The proper tooling may not be on hand, leaving workers moving back and forth to get the necessary equipment. And the station may not be set up for the job. As Boeing's CEO acknowledged, "it's uncomfortable. It creates opportunities for failure".[16] And that is exactly how the Alaskan Air aircraft ended up being delivered to the customer without the four bolts that were supposed to hold the door plug in place.[17]

Accordingly, reducing the amount of travelled work was one of the safety/quality goals to be included in Boeing's operational performance score.

To put this in a broader context, each case of travelled work is a precursor event, and reducing the number of such events reduces the risk of potentially disastrous manufacturing faults being incorporated into an aircraft. Treating the number of cases of travelled work as a KPI to be minimised clearly enhances aircraft safety.

A related concept is the idea of rework – additional work required on the assembly line because of inadequacies in the initial work done during earlier stages of production, particularly by suppliers. Rework provides additional opportunities for errors to creep in. Hence, the fewer hours that must be devoted to rework the better. Reducing the hours of rework was another goal to be included in the suit of operational performance metrics contributing to the corporate performance score, in turn a part of the annual incentive plan.[18]

Apart from contributing to the corporate performance score, these metrics have specified limits that, if exceeded, trigger corrective action or additional monitoring.

So, the situation for many thousands of Boeing employees will now be that safety/quality metrics are the most heavily weighted component of their annual incentive plan, and those metrics have been carefully chosen to enhance the quality of the assembly work and hence the safety of the finished product.[19]

A Strategy for Reducing Travelled Work

It is one thing to incentivise an improvement in a particular metric; it is another to develop a method to achieve this outcome. Consider the goal of reducing travelled work. To achieve this goal, Boeing implemented what it calls a "move ready" process. Aircraft were not to move to the next position on the factory floor until specified "build milestones" had been achieved. If it proved necessary to over-ride this requirement, a safety risk assessment would be required, and a risk mitigation plan put in place.[20] Of course, one can immediately foresee the possibility that these safety risk assessments might be carried out in a very perfunctory way, but even so, assuming the result is treated as travelled work, the incentive remains to take the build milestones seriously.

Consider, now, the "product safety downward modifier" applicable to the long-term incentives payable to top executives. One element of this modifier was the following:

- development of operational control limits on assembly lines for the 737 and other aircraft that include measures for determining when a safety risk assessment is needed before a product can move past a specified point.

This is precisely the "move ready" program just described, which aims to reduce the amount of travelled work. In 2024, top executives were to be incentivised to ensure this "move ready" program was up and running, preferably within a year, but otherwise two. A substantial portion of their long-term bonus would depend on this. This provided top executives with a powerful incentive to support Boeing's program to reduce the amount of travelled work. For the first time, Boeing had tied long-term incentive payments to ensuring aircraft safety. On the face of it, this represents a major change in the reward structure for Boeing's top executives.

Recall that the November 2021 shareholder settlement, discussed in Chapter 8, included the following requirement:

> Bonuses paid to top (named) executive officers shall include consideration of safety metrics. These will include metrics "best suited to encourage strong oversight of safety improvements and overall safety of Company products".[21]

That is exactly what Boeing did. But it went beyond this. The shareholder settlement requirement is silent on the weight to be given to these metrics, vis-à-vis the weight given to financial metrics. Boeing's 2024 incentive arrangements required not only the identification of suitable metrics but also enough weight to make a difference.

CLAWBACK

There is one other way in which, in exceptional circumstances, safety considerations may now impact on Boeing's long-term incentive payments – its clawback policy. The policy is in line with the recent Department of Justice guidance about the need for companies to make provision to "claw back" incentive payments when misconduct is discovered.[22]

Boeing's policy enables it to claw back payments made to executives who have engaged in fraud, bribery, or similar illegal acts, or knowingly failed to report such acts by an employee over whom they had direct supervisory responsibility. In addition, the Board may recover such payments from any executive who has violated company rules with respect to safety, or been negligent in supervising someone who has violated such rules.[23]

It is hard to imagine such a policy being successfully applied in the wake of the two 737 crashes or the door plug blowout. Neither fraud nor bribery seem adequate descriptions of the behaviour of senior executives prior to those accidents and nor is it obvious that anyone violated company rules with respect to safety. Moreover, as commentators have noted:

> putting these mechanisms into place will present a whole host of issues. For instance, what is the appropriate trigger for a clawback? An accusation? An adverse HR action? An indictment? A guilty plea or verdict?[24]

In the face of these difficulties, clawback appears a remote possibility in relation to the kinds of accidents under discussion. For this reason, it is hard to imagine clawback policy having much impact on the thinking of senior executives in this context. Incentives will be more effective if they are paid on an annual basis for the conscientious management of risk, as discussed above, rather than based on negative consequences when some rare event occurs. That is not an argument against clawback. The public demand for justice may in some circumstances require clawback. The only point here is that clawback is not an effective way to induce companies to focus on safety risk.

CONCLUSION

For many years, incentives for Boeings executive managers overwhelmingly emphasised the achievement of commercial goals. Although safety was said to be a top priority, it played almost no part in the determination of their bonuses. The same is true for many other large companies. That is the nature of shareholder capitalism. However, Boeing's prolonged crisis, starting with two disastrous aircraft crashes and culminating with the door plug blowout, transformed the landscape. It was now crystal clear that aircraft safety could not be taken for granted and that shareholder interests would be best served by incentivising safety in a way that had never been done before. Boeing has identified some of the precursors of poor-quality work on the assembly line and may well have effectively incentivised a reduction in the number of such events.

However, there is still a way to go. The company has yet to find ways of incentivising the quality engineering that would have avoided some of the faulty engineering decision-making implicated in the 737 MAX crashes.

Perhaps most importantly, Boeing has not effectively incentivised its executive managers to ensure that people at lower levels in the organisation are themselves incentivised to report problems before they culminate in catastrophic events. I take up this theme in Chapter 10.

NOTES

1 See, for example, Boeing's 2018 Proxy Statement, pp. 23–36.
2 Boeing 2023 Proxy Statement, p. 46.
3 "Boeing Shares Slump as Wells Fargo Cuts to Rare Bearish View", *The Seattle Times*, September 3, 2024.
4 Another qualification should be mentioned for the sake of completeness. In 2020, the CEO, but no other named executive, received a "supplemental" long-term incentive, vesting three years later, dependent on seven specific performance goals – see Boeing 2023 Proxy statement, p. 47. Most of these related to achieving business/production goals, with only one related to safety – the realignment of the engineering function. These goals had not been fully met by the end of the period and so the award did not vest. We know from the safety culture review that the realignment of the engineering function was one of the goals that was not met. Another of these goals was the achievement of crewed *Starliner* flight. This is precisely the kind of incentivised goal that puts pressure on all concerned to cut safety corners. Boeing had all kinds of problems with the Starliner and arguably the craft was launched before those problems had been resolved, resulting in the failure of its first crewed mission. "NASA decides to keep two astronauts in space until February, nixes return on troubled Boeing capsule". *Seattle Times*, August 24, 2024.
5 Boeing 2023 Proxy Statement, p. 41.
6 Boeing 2023 Proxy Statement, p. 42.
7 Boeing 2023 Proxy Statement, p.46.
8 Chapter 8.
9 Boeing 2023 Proxy Statement, p. 47.
10 Boeing 2024 Proxy Statement, p. 66.
11 Boeing 2024 Proxy Statement, p. 67.
12 For a more in depth discussion of this point, see Hopkins, A *Quiet Outrage: The Way of a Sociologist*, CCH, Sydney, 2016, Chapter 5.
13 Vaughan D, *Dead Reckoning: Air Traffic Control, System Effects and Risk.* University of Chicago Press, Chicago, 2021, p. 321.
14 Terlep, "Behind the Alaska Blowout", *The Wall Street Journal*, March 11, 2024. "F.A.A. Audit of Boeing's 737 Max Production Found Dozens of Issues" *New York Times,* March 11, 2024.
15 The following account is slightly modified from Terlep, "Behind the Alaska Blowout", *The Wall Street Journal*, March 11, 2024.
16 Terlep, "Behind the Alaska Blowout", *The Wall Street Journal*, March 11, 2024.
17 "Inside Boeing's Factory Lapses That Led to the Alaska Air Blowout", *Seattle Times*, August 25, 2024.
18 See 2024 Proxy statement, p. 66; executive summary, p. 3.
19 Operational performance metrics will also include reduction employee injury rates, but the documentation does not give any indication of weighting.

20 p. 4 of executive summary of 2024 proxy statement.

21 Exhibit A p. 3, In The Court Of Chancery Of The State Of Delaware In Re The Boeing Company Derivative Litigation)) Consol. C.A. No. 2019–0907-Mtz Stipulation And Agreement Of Compromise, Settlement, and Release. November 2021.

22 Cruz Melendez et al., "Revisions to the DOJ's Corporate Criminal Enforcement Policy…", *Skadden*, October 6, 2022.

23 Boeing Company Clawback Policy, Adopted June 27, 2023, https://www.boeing.com/content/dam/boeing/boeingdotcom/principles/ethics_and_compliance/pdf/claw-back-policy.pdf

24 See Skadden advice referenced above, p. 3.

10 Reflections on Safety Culture

"Safety culture is arguably the single most important influence on the management of safety".

So says the International Civil Aviation Organization (ICAO).[1] It is a striking claim which, if accepted, requires that we think carefully about the meaning of safety culture and how it functions.

It is also a controversial claim. Judge Moshansky, who led one of the most far-reaching inquiries ever into aviation safety, after the crash of an Air Ontario passenger plane in 1989 in which 29 people died, makes the following claim:

> [A] Safety Management System, if it is to succeed, … must be accompanied, … by an effective, properly financed and adequately staffed system of monitoring, surveillance and enforcement on the part of the Regulator.[2]

It is odd that ICAO, headquartered in Quebec, appears unaware of this competing, even contradictory, claim. I return to this at the end of this chapter.[3]

But let us accept at face value the ICAO claim that safety culture is the critical issue and see where it gets us. It is all the more important to consider it carefully because establishing a positive safety culture is often assumed to be the key to getting Boeing back on track.

THE LEGISLATIVE BASIS

Perhaps a good place to start is a 2020 Act of the U.S. Congress, with the awkward title – the Aircraft Certification, Safety, and Accountability Act.[4] This was Congress's legislative response to the 737 MAX disasters, but it was a rushed reaction to the crisis of public confidence in Boeing and was not well considered. I shall explain this comment shortly, but right now, here is one striking instance of the problem.

The Act requires that Boeing, and similarly situated companies, adopt a code of ethics "which clarifies that safety is the organization's highest priority".[5] There is something absurd about this requirement. As we saw in Chapter 9, until very recently, Boeing's top corporate office holders were remunerated overwhelmingly on the basis of the profitability of the company. They were incentivised to put shareholder returns above all else. Where such a situation prevails, there is no way that safety can be said to be the organisation's highest priority, regardless of any *ethical clarification* to the contrary. What Congress really needs to do is look closely at these remuneration systems and find ways to intervene.

But returning to the Act, probably its most important provision was that companies to which the FAA delegates regulatory/certificatory functions, like Boeing, must develop "a safety management system", as recommended by ICAO.[6] A safety

DOI: 10.1201/9781003629269-10

management system goes beyond the detailed prescriptive rules with which safety regulators typically operate in the U.S. It requires, in addition, that companies proactively identify and manage all the hazards they create. There is an instructive precedent for this. After the Deepwater Horizon accident in the Gulf of Mexico in 2010, the regulator made it mandatory for operators to have a safety and environmental management system. This was something the industry had long resisted, but Deepwater Horizon highlighted the shortcomings of prescriptive regulation in such a dramatic way that industry had little option but to accept the new requirement.[7]

However, it is a second provision in the Act that is the subject of this chapter. This provision in effect mandates that companies like Boeing must work to achieve a positive safety culture. I say "in effect" because the provision did not specifically require companies to have a positive safety culture. Instead, it set up an Expert Panel to examine the extent to which Boeing (and similarly situated companies) had implemented "a safety culture consistent with the principles of the ICAO's safety management manual".[8]The Panel's role was to make recommendations to FAA based on its findings. The FAA would then be expected to implement these recommendations as best it could.[9]

Using legislation to mandate a safety culture is fraught with difficulties that Congress seems to have passed over. The Norwegian experience is worth recounting. In 2002, Norway introduced regulations requiring companies in the offshore oil and gas industry to develop a "good and sound HSE culture"[10] (HSE refers to Health, Safety, and Environment, but for simplicity, I treat this as synonymous with safety culture). The regulations did not define what a safety culture was, which led to a great deal of discussion and debate as to precisely what regulators and companies needed to do. How could regulators enforce this new requirement if there was no clarity about what it really meant? Was safety culture just another label for more concrete activities and if so, did this new label add value? Was a good safety culture just another way of talking about quality risk management, and therefore tautologous? In the end, these questions proved insurmountable, and the industry and its regulator lost interest, moving on to more tangible safety issues.

THE REASON MODEL OF SAFETY CULTURE

All of this raises the question of where the concept of safety culture came from in the first place. It made its appearance in the safety literature in the 1980s but gained real prominence when a U.K. scholar, James Reason, wrote about it in his influential book, *Managing the Risk of Organizational Accidents*. The Norwegian attempt to legislate this idea was heavily influenced by Reason and his book. I mention this here because Reason's name will crop up more than once in this account.

So what was Congress referring to when it spoke of safety culture? The Act refers specifically to the principles outlined in the ICAO safety management manual. Chapter 3 of the manual is devoted to this topic. Unfortunately, it is quite inadequate. Despite stating that "safety culture is arguably the single most important influence on the management of safety", nowhere does it provide a clear definition of safety culture. The closest it comes to it is this:

> Safety culture has been described as "how people behave in relation to safety and risk when no one is watching."[11]

ICAO doesn't specifically endorse this definition, which is just as well, since it reduces culture to a characteristic of individuals. In contrast, Reason, and many other scholars, treats safety culture as an organisational property. For instance, at one point Reason writes:

> This chapter [emphasises] the critical importance of an effective safety information system [in defining a good safety culture].[12]

Obviously, an effective safety information system is an organisational characteristic, not a characteristic of individuals.

ICAO's failure to define safety culture clearly needs to be put in context. Despite a voluminous academic literature on the subject, no agreed definition of the term has emerged, and scholars are increasingly suggesting that the term should be abandoned. This is not the place for an extended discussion but interested readers are referred to various writings on the subject.[13]

It must also be noted that, while the Act requires the Expert Panel to take note of the principles referred to in the ICAO document, there are no explicitly stated safety culture principles in that document, and it is difficult even to infer such principles. So what was the Panel to do? It chose to base itself on the work of James Reason.[14]

Reason proposes a model of safety culture that includes five components that collectively make up a good safety culture. They are:

- a reporting culture,
- a just culture,
- a flexible culture,
- a learning culture, and
- an informed culture.[15]

These five components are used by many government agencies, including NASA and the FAA, and by Boeing itself. Somewhat inexplicably, the ICAO document fails to refer explicitly to these five components, even though ICAO has been strongly influenced by Reason's work.[16] But given that Boeing uses the Reason framework, this was a good basis on which to evaluate Boeing's safety culture.

The Expert Panel made various observations on Boeing's safety culture as a whole, based on extensive interviews and a study of the documents. It noted that Boeing had employed a safety culture "expert" to champion Boeing's safety culture work. But it noted that Boeing's implementation of the five components of a positive safety culture has been "inadequate and confusing".

Here is one example that makes the point. One of the five components of a safety culture identified by Reason is that it be "informed". Boeing sees this as meaning it is the responsibility of each employee to complete their own training, and the manager's responsibility to ensure they can get the training they need.[17] This is a long way from Reason's idea that the basis of an informed culture is an effective safety information system that "collects, analyses, and disseminates information from incidents and near misses as well as from regular proactive checks".[18] To reduce the idea of an informed culture to one in which individuals are held responsible for completing their training is a travesty of the original idea.

The Panel also concluded that "employees are confused by the different terms and the lack of explicit descriptions that clarify these topics".[19] This is hardly surprising, given the uncertainty about exactly what a safety culture is. It also raises the important question of why employees need to be trained on the topic of safety culture at all. Scholars may argue about the meaning of safety culture, but there is no reason to require employees to understand the meaning of the term. If Boeing regards a reporting culture as important, then employees need to be rewarded for reporting; if Boeing is seeking a just culture, then it must convince its employees that they will not be blamed if they report. Arguably, safety will be better served by by-passing altogether the idea of safety culture and dealing with the five components separately, and even better, narrowing the focus to particular, more concrete components, such as reporting. The academic literature is now beginning to suggest this disaggregated approach as a way forward.[20]

The folly of seeking to create a safety culture simply by training workers in the theory of safety culture has been repeatedly demonstrated. Years ago, the energy company, Shell, decided to train its entire workforce of 250,000 people in a variant of Reason's safety culture model – the organisational maturity model. This model envisages a ladder of organisational culture types, ranging from a pathological culture (who cares, as long as we don't get caught) through to an ideal safety culture[21] (safety is how we do business around here). It was hoped that workers would recognise the type of culture that best described their own organisational context and begin to climb the ladder towards the ideal safety culture. But despite this massive educational effort, there was no discernible improvement in safety. The workforce had learned the language of the organisational maturity model, but Shell had not introduced the kinds of organisational changes necessary to support it, in particular, a system of recognition and rewards to encourage the required behaviour.[22]

NARROWING THE FOCUS

Returning to the disaggregated model of a safety culture, which of the five components listed above should be the focus of attention? This question is answered most clearly in the literature on High-Reliability Organisations (HROs), which closely parallels the safety culture literature, dealing with many of the same issues, although using somewhat different terminology. Here is what one of the originators of the HRO tradition writes:

> The key difference between HROs and other organisations in managing the unexpected often occurs at the earliest stages when the unexpected may give off only weak signals of trouble. The overwhelming tendency [of most organisations] is to respond to weak signals with a weak response. ...[HROs] see the significant meaning of weak signals and give strong responses to weak signals.[23]

Giving a strong response to weak signals is probably the best known and defining characteristic of HROs. Clearly, responding to weak signals requires highly developed reporting systems, that is, a highly developed culture of reporting. This is arguably the paramount feature of an HRO. Analogously, a reporting culture is the paramount feature of a good safety culture. A close reading of Reason's text, particularly on page 195, reveals that this is indeed his view.

Reason notes further that a willingness to report, even when it involves one's own errors or near misses, depends on a clear understanding that one will not be blamed or punished. On the contrary, the expectation is that one will be thanked or even rewarded for bringing such matters to attention. In short, a reporting culture needs to be a no-blame culture in order to flourish. At this point, critics will point out that a blanket amnesty for all unsafe acts would be unworkable and would indeed lack credibility in the eyes of the workforce. True. That is why, overall, organisations need to implement, not a no-blame culture, but a just culture, where some behaviour is deemed unacceptable and blameworthy. However, for a reporting culture to flourish, reporters need to feel confident that they will not be blamed for anything they report. If there is any doubt about this at all, they will simply not report. The concept of just culture must always be articulated and promoted with that proviso clearly in mind.

To summarise, the safety literature is clear that a good reporting culture and a just culture – which must be a no-blame culture in relation to reporting – are the two most important ingredients of a good safety culture.

These two components were highlighted by the Expert Panel.[24] Moreover, according to the Panel, Boeing itself seems to have focused its safety culture implementation efforts on Just Culture and Reporting Culture.[25]

BOEING'S REPORTING SYSTEMS

So how successful has Boeing been in implementing a no-blame reporting culture, according to the Expert Panel?

In 2020, that is, after the two crashes, Boeing introduced a programme to encourage "Seek, Speak, and Listen" behaviours. However, according to the Panel, the implementation of this programme involved a good deal of speaking [presumably by management] with little or no attention paid to seeking or listening.[26]

Boeing also operates "an anonymous reporting programme, "Speak Up", which is its preferred channel for employee reports. However, the Panel found that employees distrust the anonymity of the programme, which undermines its effectiveness. They preferred to report to their immediate supervisors, which unfortunately meant that these reports were more readily discounted. It also meant that such reports were not consistently captured in Boeing's safety management system. Finally, employees were not consistently given feedback. Clearly, a lack of feedback will predictably discourage further reporting.

In short, the Panel found that Boeing had failed to create a culture of reporting, in part because it had not convinced employees that it was safe to make reports.

Boeing claims to have taken action, since the door panel blowout of January 2024, to improve the effectiveness of employee reporting channels. During hearings conducted by the National Transportation Safety Board (NTSB) in August 2024, it claimed that it had received well over 2,000 reports into the Speak Up system since the beginning of the year. However, fear of retaliation was still widespread. The head of the NTSB spoke at the hearings of the case of two employees who had been involved in the failure to replace the bolts on the door plug that blew out. There was no evidence that their mistake had been intentional, she said, yet Boeing had placed them on administrative leave in a way that was widely understood as punitive. It was clear that Boeing had failed to convince her that it operated a just culture.[27]

Nor is Congress convinced that Boeing has either a reporting culture or a just culture. The 2020 Act contains explicit protection for whistleblowers who make reports to the federal government or to the employer.[28] The need for whistleblower protection is an implicit recognition that reports may be unwelcome and that reporters run the risk of retaliation. The legislation is an acknowledgement that Boeing is a long way from the safety culture to which it aspires.

To summarise, Congress and other parties have concluded that a defective safety culture was one of the causes of Boeing's troubles. But the concept of safety culture is so confused that perhaps we should abandon it and focus instead on its components, in particular a reporting culture and a just culture. Boeing fails on both these counts.

HOW TO CREATE A CULTURE OF REPORTING

The aim of a reporting system must be to identify the "bad news" that is usually available on the shop floor or in engineering offices, so that something can be done about it before it culminates in a damaging event. Here are some principles, drawn from the literature on HROs, about how to institutionalise an effective culture of reporting in hazardous organisations.[29]

PRINCIPLES FOR CREATING AN EFFECTIVE "BAD NEWS" REPORTING CULTURE

- The reporting technology must be as user-friendly as possible.
- Reports should be routed automatically to various designated people.
- All reports should be individually responded to.
- Companies must encourage and reward "helpful" reports.
- Companies must encourage and reward courageous reports.
- The system must be no-blame.
- Contract workers must be encouraged to participate in the client's reporting system.
- There should be no reporting targets.
- A bad news reporting system depends crucially on top organisational commitment.
- Governments must guarantee bad news reporting systems will not increase the risk of prosecution.

The end result of a bad news reporting system that is working well is a highly risk-aware workforce. Employees are alert to warnings of danger and to precursors of all sorts. They exhibit the type of mindfulness that is characteristic of HROs. They do so because the system is one that encourages, recognises, and rewards this mindset. It continually reinforces risk-awareness by identifying reports that are most helpful in reducing fatality risk.

MOSHANSKY'S CLAIM

Finally, let us return to Judge Moshansky's claim that an effective safety management system depends on a well-resourced and resolute regulator; he makes no mention of a positive safety culture. What are we to make of this discrepancy?

Moshanky's final report, appeared in 1992, three years after the Air Ontario crash. This was several years before the concept of safety culture came into widespread use. Could it be that if the concept of safety culture had been available when he was writing his report, he would have seen Air Ontario's inadequate safety culture as the cause of the crash, or at least a contributing factor? This is hardly plausible. Here again are his words:

> The lack of adequate funding of Transport Canada's Regulatory Oversight branch in the 1980s was the root cause of the Dryden Crash[30]

In his report, Moshansky had recommended a safety management system (SMS) initiative, "with the proviso that such a scheme must be accompanied by a strong Regulatory Oversight regime". As it turned out, Transport Canada began introducing the SMS initiative in 2005, not as a way of improving safety, but as a way of shedding its own responsibility for safety, forced on it by government funding cuts. The result was a new SMS requirement without the regulatory oversight that Moshanky deemed essential. It is hard to see that the concept of a safety culture, positive or negative would have added anything to his analysis. Rather I think we should take from Moshansky the idea that an effective safety management system, which includes a robust, no-blame reporting culture, is dependent on strong regulatory oversight. That is also one of the paramount lessons arising from the 737 MAX tragedies.

NOTES

1 ICAO – Safety Management System – Doc 9859 – 4th ed., 2018, pp. 3-1.
2 Moshansky V. *Brief presented at hearings of the House of Commons Standing Committee of Transport, Infrastructure and Communities*, Ottawa, Thursday, April 6, 2017.
3 There is also evidence that the direction of influence is the other way around from that claimed by the ICAO – that an effective safety management system must be in place before it is possible to start developing a good safety culture. (Naevestad T and Phillips R, "The Limits of Soft Safety Regulation: Does Successful Work with Safety Culture Require SMS Implementation", *Transportation Research Interdisciplinary Perspectives*, 17 (2023).
4 Aircraft Certification, Safety, and Accountability Act of 2020 (Pub. L. 116–260, Div. V).
5 S. 102f.

6 ICAO was created in 1944 "to promote the safe and orderly development of civil aviation around the world. The organization sets standards and regulations necessary for aviation safety, security, efficiency, and regularity, as well as for aviation environmental protection."

7 See my *Disastrous Decisions: The Human and Organisational Causes of the Gulf of Mexico Blowout*. CCH, Sydney, 2012, Chapter 10.

8 S. 103(a)2A.

9 By way of clarification, the committee chose to focus on Boeing and assumed that its recommendations would assist other similarly situation companies. Expert Panel Report, p. 5.

10 Kringen J. "Contested terrains in risk regulation: Legitimacy challenges in implementation process", in Lindoe, P, Baram, M, Renn O, (eds) *Risk Governance of Offshore Oil and Gas Operations,* Cambridge University Press, 2014.

11 ICAO Safety Manual, chapter 3-1 https://www.icao.int/SAM/Documents/2017-SSP-GUY/Doc%209859%20SMM%20Third%20edition%20en.pdf

12 Reason J, *Managing the Risk of Organizational Accidents*, Ashgate, 1997, p. 194.

13 https://www.ohsbok.org.au/chapter-10-2-1-organisational-culture-a-search-for-meaning/

14 https://www.faa.gov/regulationspolicies/rulemaking/committees/documents/section-103-organization-designation. Hereafter, EP, p. 9.

15 Reason J, *Managing the Risk of Organizational Accidents*, Ashgate, 1997, pp. 195–196.

16 See Section 2.3 of International Civil Aviation Organization, *Safety Management Manual*, Fourth Edition (ICAO Doc. No. 9589).

17 EP, p. 23.

18 Reason J, *Managing the Risk of Organizational Accidents*, Ashgate, 1997, pp. 194,195.

19 EP, p. 22.

20 "Seeking a Scientific and Pragmatic Approach to Safety Culture in the North American Construction Industry", *Safety Science*, forthcoming.

21 This was given the confusing title – a generative culture.

22 For a more detailed account see Hopkins, *Organising for Safety: How Structure Creates Culture*, Wolters-Kluwer, Sydney, 2019, pp. 31–34.

23 Weick K and Sutcliffe K, *Managing the Unexpected: Assuring High Performance in an Age of Complexity* Jossey Bass, San Francisco, 2001, pp. 3–4.

24 EP, p. 4.

25 EP, p. 21.

26 EP, pp. 21, 22.

27 "NTSB Chair: Boeing Retaliated against 2 Workers on the 737 that Blew Out", *Seattle Times*, August 7, 2024. https://www.seattletimes.com/business/ntsb-chair-boeing-retaliated-against-2-workers-on-737-that-blew-out/#:~:text=National%20Transportation%20Safety%20Board%20Chair,it%20will%20not%20take%20disciplinary

28 S. 118.

29 See Hopkins A, "A practical guide to becoming a high reliability organisation", http://hdl.handle.net/1885/287366

30 Mosh Moshansky V. *Brief presented at hearings of the House of Commons Standing Committee of Transport, Infrastructure and Communities*, Ottawa, Thursday, April 6, 2017, p. 2.

11 The Causes of the 737 MAX Crashes

What caused the 737 MAX crashes? This is a natural question to ask. But it is also deeply problematic. The problem stems from the word "cause" itself, which is deceptively simple, but turns out to have multiple meanings and connotations. Accident analyses often skip over this problem and treat the word as an undefined concept. In particular, many so-called "root cause" investigations are carried out without ever defining this term, even though the idea of a single *root* cause is even more problematic than the idea of cause itself.

AN ACCIMAP OF THE ETHIOPIAN AIRLINE CRASH

The best way I know to approach this question of causation is to develop a diagram known as an accimap ("acci" pronounced as in "accident"). This technique was introduced into the safety literature by a Danish accident analyst, Jens Rasmussen.[1] Accimaps provide a visual summary of the arguments made in the text.

Figure 11.1 is an accimap. It brings together nearly all the factors mentioned in the book as contributing to the Ethiopian Airlines accident. A very similar diagram could be drawn for the Lion Air accident and a somewhat similar diagram could be developed for the door plug blowout. Accimaps locate the contributory factors in terms of their organisational and temporal remoteness from the accident. The categorisation of these factors can vary, depending on the information available. In this map, there are three categories or groupings of contributory factors, listed on the left side: background, specific to the MAX, and on the day. Figure 11.2 is an index to the chapters in which the factors are discussed, so interested readers can verify details for themselves.

The meaning of the arrows can also vary. In this map, a meaning has been chosen which allows us to speak of the factors in the diagram as causes of a particular kind. A → B means if A had not occurred or had been absent, B would not have occurred. This is causation in the but-for sense: but-for A, B would not have occurred. For example, but-for the decision to re-engine the 737, the chain of events leading to the accident would never have occurred. This is undeniably true.

Identifying but-for causes enables us to make deterministic statements, not probabilistic ones. So, had the MCAS been designed to be error-tolerant, the accident would not have happened; if the MCAS software had been connected to two angle-of-attack sensors, instead of only one, the accident would not have happened. Likewise, the accident would not have happened if the FAA had not been misled about the MCAS, if the Airbus challenge had not triggered a decision to re-engine the plane, and if Boeing had not been transformed in the late 1990s into a business

DOI: 10.1201/9781003629269-11

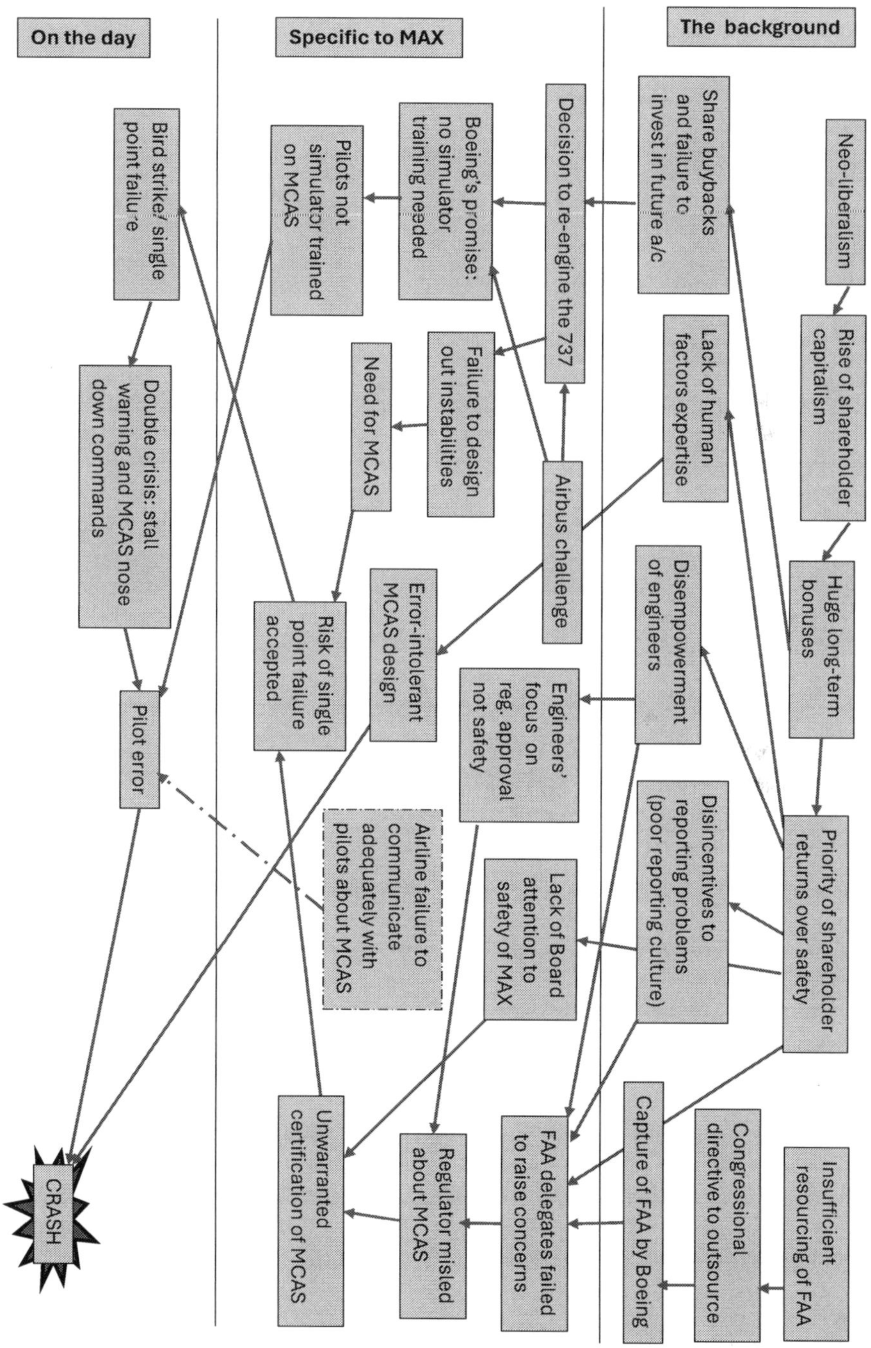

FIGURE 11.1 Map of factors contributing to the crash of Ethiopian aircraft.

Airbus challenge	Ch4	Huge long-term bonuses	Ch4
Airline failure to com...	Ch6	Insufficient resources	Ch7
Bird strike	Ch6	Lack of board attention	Ch8
Boeing promise	Ch4	Lack of human factors exp...	Ch6
Capture of FAA	Ch7	Need for MCAS	Ch5
Congress directive	Ch7	Neo-liberalism	Ch3
Crash	Ch2	Pilot error	Ch6
Decision to re-engine	Ch4	Pilots not trained	Ch5
Disempowerment of engineers	Ch3	Priority of shareholders	Chs 4,9,11
Disincentive to reporting	Ch10	Rise of shareholder cap	Ch3
Double crisis	Ch6	Risk of single point failure	Ch5
Engineers focus on reg approval...	Ch7	Regulator misled	Ch7
Error-intolerant MCAS	Ch6	Share buybacks	Ch4
FAA delegates failed to raise....	Ch7	Unwarranted certification	Ch7
Failure to design out hazards	Ch5		

FIGURE 11.2 Index of chapters in which factors are discussed.

in which shareholder interests trumped all other considerations. The truth of each of these statements is established in the text.

The diagram is necessarily complex. But it is worth spending time tracing some of these pathways. For instance, the claim that neo-liberalism was a but-for cause of the crash of the Ethiopian aircraft may at first sight appear outlandish. But it will appear considerably more reasonable if one follows some of the pathways from the top left to the bottom right of the accimap.

The wording in the boxes is necessarily much abbreviated and may cover several ideas. For example, "disincentives to reporting problems (poor reporting culture)" covers the treatment of whistleblowers, while reporting culture is the clarification of safety culture, as discussed in Chapter 10.

One of the contributory factors shown on the accimap – "airline failure to communicate adequately with pilots about MCAS" – has a broken border and is connected to "pilot error" by a broken arrow. This is intended to indicate a divergence of opinion: the French report regarded this as a contributory factor; the Ethiopian report did not.[2]

One of the benefits of an accimap representation of an investigation is that it lays out systematically the possible points of intervention uncovered in the inquiry. In other words, it identifies possible levers of change. We can then ask: who has their hands on these levers? Governments, regulators, corporate boards, remuneration consultants, design engineers, chief engineers, unions and professional associations, safety managers, and procedure writers – these are some of the parties with the ability to pull particular levers of change. All such parties can look at an accimap and decide whether it suggests courses of action for them.

But-for causes do not lead inevitably to a particular final outcome – they are necessary, but not sufficient in themselves to result in that outcome. Rather, they make the particular outcome possible; in their absence that outcome would not occur.

It is perhaps worth pausing to note that no single but-for cause can ever be sufficient to cause an accident; all that can be said about it is that it was one of a potentially infinite number of factors that together were sufficient to cause the accident.[3] Let me develop this point a little further. Simple-minded approaches to accident causation sometimes identify an error or violation by a front-line worker as *the* cause of the accident, which implicitly means the error or violation was sufficient to cause the accident. Was pilot error, perhaps, sufficient to cause the Ethiopian Airlines crash, as Boeing initially claimed? At the bottom right of Figure 11.1, two factors lead directly to the crash – pilot error *and* an error-intolerant MCAS design. The fact is neither one of these was sufficient. In particular, consider pilot error. If the MCAS design had been more error-tolerant, the pilots would almost certainly have recovered from their errors. Recall that the pilots managed to counteract the first nose-down instruction given by MCAS. The error-tolerant MCAS that Boeing subsequently devised would not have initiated the second and later nose-downs. So, after the first nose-down recovery, the pilots would have been free to deal with the false stall alarm. In short, pilot error did not by itself lead inevitably to the accident; nor did the error-intolerant MCAS design. Both were necessary.

Could it be that pilot error *plus* an error-intolerant MCAS were together sufficient to cause the accident? I made reference in Chapter 2 to a stroke of luck that prevented pilot error and an error-intolerant MCAS from causing the Lion Air plane to crash on the sector before the one on which it in fact crashed. The stroke of luck was that there happened to be a third, off-duty pilot sitting in a jump seat in the cockpit.[4] He had the luxury of observing what was happening and realising what needed to be done. Had he not been present, the aircraft would probably have crashed in that sector rather than the next. The absence of a third pilot in the cockpit of the Ethiopian aircraft is therefore a third factor that needs to be included in any attempt to identify a sufficient set of causes. Clearly, the list can be extended indefinitely.[5]

Some accimap analysts are not comfortable with the deterministic nature of but-for causation and prefer to interpret the arrows on their diagrams in probabilistic term, that is, A → B means A increased the likelihood of B. The accimap then consists of a network of these probabilistic connections. This makes it possible to include contributory factors even where they do not satisfy the but-for logic. However, there is a serious drawback to this approach. The further back one moves along a chain of causation, each link of which has an uncertain probability, the less likely it is that a remote causal factor of interest had any significant impact on the final outcome. Such accimaps are still useful for thinking about possible points where safety improvements can be made, but they are less than adequate as explanations for a particular accident.

An accimap is not a tool that can be used to carry out an investigation. It is a convenient way of summarising the *results* of an investigation that has already been conducted. Nor, by itself, does it provide the argument or evidence necessary to justify the arrows. Arrows can only be included on an accimap if the argument connecting two boxes is made in the accompanying text.

Accimaps from other investigations are of value in that they can be used to guide the search for contributory factors for a current investigation, but the shape and details of an accimap for the current investigation can only be determined after the investigation is complete. Moreover, different investigators will highlight different factors in their investigations of the same accident and their accimaps may look significantly different. At this stage, construction of accimap is more an art than a science.

Accimaps can be read as the results of a why-tree analysis, starting at the bottom. So, why did the Ethiopian Airlines pilots fail to take effective action to deal with the crisis they were confronted with? The accimap immediately identifies various but-for causes, which of course counteracts the impulse to blame the pilots. The entire structure of the map works in this way to eliminate blame. Take the box – "regulator misled about MCAS". This is a reference to the Boeing technical pilots, who misled the FAA. They did mislead the FAA, in that they failed to explain to the FAA that MCAS was vulnerable to a single-point failure, but there were various reasons for this failure, including that they were unaware of this vulnerability, as discussed in Chapter 8. To blame them for the crashes as the Department of Justice tried to do is quite inappropriate, given the web of causation in which they were embedded. The same is true, even for the CEO and Board, who themselves are subject to the forces of shareholder capitalism. This does not mean we should ignore the contribution of the CEO and Board to an accident such as the crash of the Ethiopian aircraft, as I shall now explain.

FINAL REFLECTIONS

Consider the accimap box "priority of shareholder returns over safety", on the top line of the diagram. The accimap suggests that this factor contributed to five others. If we could change this one cause, there would clearly be a considerable knock-on effect. If this factor had been otherwise, much of the contributory network would have fallen apart. So how might it have been otherwise? Consider the previous box, "huge long-term bonuses" which, as explained earlier, were based exclusively on financial performance. If these long-term bonuses had given priority to product safety, we can be fairly confident that this accident would not have happened. It is for the Board to determine the details of long-term bonuses and, as we saw in Chapter 9, the Board did introduce some significant product safety goals in 2024 which, if not met, would result in the cancellation of a significant part of the CEO's long-term bonus. This was a small, but promising, step towards the goal of a more general reversal of the priority of shareholder returns over safety.

There are other ways of counteracting unrestrained shareholder capitalism, ways that are available to governments. One such intervention would be to introduce an *effective* legal requirement, backed by criminal sanctions, that CEOs and individual board members exercise *reasonable care* for the safety of their product. A key word here is effective. Such legal requirements exist in many jurisdictions,[6] but they are often not effectively enforced. It is not being suggested here that top corporate officers be punished because they are culpable or blameworthy – arguably they are no more culpable than anyone else whose behaviour contributed to the accident. The argument is that if there is a realistic possibility that failure to exercise reasonable care might render one criminally liable, this has the potential to encourage the necessary focus on safety, despite all the financial pressures to the contrary. The use of criminal sanctions in this way is based on consequentialist considerations, not on any notion of desert or blame.

What might "reasonable care" mean for a company like Boeing? The shareholder judgment discussed in Chapter 8 provides a detailed guide. If the criminal law were to follow the civil law in this respect, this would be a significant new influence on the behaviour of Boards and CEOs of companies like Boeing.

The criminal prosecution of Boeing for the alleged fraudulent behaviour of two mid-level employees was inappropriate for at least two reasons: first, it focussed on the wrong people, and second, the term fraud is hardly an adequate description of the behaviour in question. On the other hand, a successful prosecution of senior corporate officers for failure to exercise reasonable care would make a lot more sense and would be likely to have a salutary effect on Boards and CEOs more generally. Such prosecutions would strengthen the restraints on shareholder capitalism. The time has surely come to return to some form of stakeholder capitalism.

NOTES

1 Rasmussen J, "Risk Management in a Dynamic Society: A Modelling Problem", *Safety Science*, 27 (2/3), pp. 183–213, 1997. See also Branford K, et al., "Guidelines for Accimap Analysis", Chapter 10 in Hopkins A (ed) *Learning from High Reliability Organisations*. CCH, Sydney, 2009.
2 An argument can be made that the Ethiopian view on this is preferable, but it would take us too far afield to develop it here.
3 Fischer D, "Insufficient Causes", *Kentucky Law Journal*, 94(1), pp. 67–107.
4 Robison P, *Flying Blind*, Penguin Random House, New York, 2021, p. 272.
5 For a further discussion of necessary and sufficient causes, see my "Leveson and Dekker on Reason: How the Critics Got the Swiss Cheese Model Wrong", https://lnkd.in/g3DzTv93. Scroll down to file near bottom of page.
6 See e.g., Tooma M, *Duty of Officers, 3rd Edition,* Wolters Kluwer, Sydney, 2021.

Made in the USA
Monee, IL
13 September 2026

62977121R10052